SMALL FARM
HANDBOOK

Project Coordinators
Shirley Humphrey, Staff Research Associate, Small Farm Center, University of California, Davis
Eric Mussen, Specialist, Entomology Department, University of California, Davis

Editors
Shirley Humphrey, Staff Research Associate, Small Farm Center, University of California, Davis
Eric Mussen, Specialist, Entomology Department, University of California, Davis
Claudia Myers, Associate Director, Small Farm Center, University of California, Davis
Ronald E. Voss, Director, Small Farm Program, University of California, Davis
Christie Wyman, Writer-Editor, Davis, California

Small Farm Program
University of California
Division of Agriculture and Natural Resources
Publication SFP001

ORDERING
For information about ordering this publication, write to
Publications
Division of Agriculture and Natural Resources
University of California
6701 San Pablo Avenue
Oakland, CA 94608-1239

Within California, call toll-free (800) 994-8849
telephone (510) 642-2431
FAX (510) 643-5470

Publication SFP001

ISBN 1-879906-23-6
Library of Congress Catalog Card No. 94-61620
©1994 by the Regents of the University of California

Printed in the United States of America

To simplify information, trade names of products have been used. No endorsement of named or illustrated products is intended, nor is criticism implied of similar products that are not mentioned or illustrated.

5m-10/94

CONTENTS

INTRODUCTION

Small-scale farm operators are transforming American agriculture by their innovative selection and careful nurturing of crops and their persistent marketing techniques. They provide most of the specialty products that so many of us have come to expect—organic produce, ethnic produce, unusual and heirloom varieties.

Their greatest successes are in locating niche markets and developing new products to satisfy them. In a city area predominantly populated by Armenians a market is in place for grape leaves, olive oil, and other products needed for Armenian cuisine. The farmer who can fill those needs already has his market. The market for many different kinds of ethnic food is increasing as immigrant populations increase.

Innovative farmers grow specialty fruits, vegetables, nuts, berries, herbs, Christmas trees, flowers, and turf grass; tend bees; raise elk, emu, salmon, or other unusual livestock, poultry, or seafood. Small farmers produce heirloom and novel varieties in small quantities to sell directly to consumers. Some of these specialty crops are not suitable for conventional wholesale markets because of their perishability. Unusual produce, such as white apricots, yellow tomatoes, Italian agretti, caprifigs and plumcots sell well after consumers get used to them. There is increasing consumer demand for "organic" produce.

Growing in greenhouses enables farmers to harvest and sell their products out of season, which makes them more desirable (and expensive) to consumers. Some farmers extract extra income from conventional products by adding value. Examples include processing raw products into pestos, jams, jellies, and sauces, producing floral arrangements, or drying and packaging herbs. Offering services to other farmers, neighbors, and tourists, including equipment rental or repair, food processing, recreation, forestry, and other enterprises based on natural resources are money making opportunities for small-scale farmers.

Diversifying farming operations creates a greater chance for year-around income. Farmers diversify their crop mixes, use trees with marketable crops for windbreaks, and produce small amounts of very labor-intensive-but-high-value crops such as European melons, figs, or herbs.

One exciting idea is community supported agriculture. This business arrangement between farmers and their urban neighbors assures the farmer of a market and the consumers of a steady supply of produce or other farm products. The arrangement also familiarizes urbanites with agriculture, thereby decreasing conflict between rural and urban land use.

Small-scale farm operators sell through many outlets: cooperatives; farmers' markets; distributors, brokers, processors; to restaurants; through the mail; and on their farms at roadside stands and pick-your-own operations.

Small-scale farm operators do not fit into a mold. Many farm part-time and bring in extra money with off-farm jobs. Some live in the city and travel out to farm relatively small plots of land. Others are urban residents who produce their crops in the city on small plots of land. Some rural residents enjoy country life more than farming, but produce some commodities. Sons and daughters farm part-time while they go to school or work at other jobs. Retirees buy a few acres to grow specialty vegetables or berries.

When they are ready to retire, farmers without heirs or family members interested in farming sometimes face a dilemma: should they sell to the highest bidder (likely a developer) and have a steady income or should they try to preserve the land as farmland. The Williamson Act helps these farmers meet both their needs. Also, there are several programs that help retiring farmers sell to beginning farmers who will help preserve the rural community; terms can be worked out that will be economically profitable to both.

Farmers are continuing to increase their efforts to work with nature and operate within its carefully balanced biological cycle. Livestock enhance vegetable farming operations by keeping fields clear, producing manure, controlling pests, and providing food. Many farmers are incorporating alternative farming methods into their operations. Sustainable, organic, biodynamic, and holistic resource farming are economically attractive, contaminate less, and conserve more natural resources for the future.

Farmers are not just tillers of the soil. They are business people, innovators, and land stewards. They need reams of information to manage their farms properly and profitably. Prospective, new, and experienced farmers who want to diversify their farming enterprises need information on marketing, agronomy, pest control, regulations, production, taxes, etc. Those who develop written business plans have greater chances of success. The uncommitted entrepreneur will have a hard time succeeding. This book explains, with anecdotal narratives and worksheets, the many factors important to small-scale farm operators.

There are many as yet unperceived needs that farmers can meet in the future. The opportunities are exciting.

—*Shirley Humphrey and Eric Mussen*

AUTHORS

Patricia Allen, Administrative Analyst, Center for Agroecology and Sustainable Food Systems, UC Santa Cruz

Aziz Baameur, Farm Advisor, Riverside County Cooperative Extension

Gregory Encina Billikopf, Farm Advisor, Stanislaus County Cooperative Extension

Steven Blank, Extension Economist, Agricultural Economics, UC Davis

Daniel W. Block, Professor, California State Polytechnic University, San Luis Obispo

Stephen Brown, Farm Advisor, Ft. Meyers, Florida, Cooperative Extension

William Burrows, Instructor, Shasta Community College

Mark Casale, Chef, Dos Coyotes Restaurant, Sacramento

Fred Conte, Extension Aquaculture Specialist, Department of Animal Science, UC Davis

Stephanie Coughlin, Farmer, San Diego County

Ralph Ernst, Extension Poultry Specialist, Department of Avian Sciences, UC Davis

James W. Garthe, Department of Agricultural Engineering, Pennsylvania State University

John M. Harper, Livestock and Natural Resources Advisor, Mendocino County Cooperative Extension

Shirley Humphrey, Staff Research Associate, Small Farm Center, UC Davis

Pedro Ilic, Farm Advisor, Fresno County Cooperative Extension

Adel A. Kader, Professor, Department of Pomology, UC Davis

Howard W. Kerr, Jr., USDA, CSRS, Office for Small Scale Agriculture

Lisa Kitinoja, Consultant, Extension Systems International, Woodland

Karen Klonsky, Extension Specialist, Agricultural Economics Department, UC Davis

Art Lange, Farmer, Honey Crisp Farms, Reedley

Pete Livingston, Staff Research Associate, UC Davis

Amy Lyman, Organizational Consultant, Great Place to Work, San Francisco

Jeanne McCormack, Farmer, Rio Vista Area

Faustino Muñoz, Farm Advisor, San Diego County Cooperative Extension

Eric Mussen, Extension Apiculturist, Entomology Extension, UC Davis

Claudia Myers, Associate Director, Small Farm Center, UC Davis

Aaron O. Nelson, Farm Advisor, Fresno County Cooperative Extension

Wynette Sills, Farmer, Pleasant Grove Farms, Sutter County

Richard Smith, Farm Advisor, San Benito, Monterey, and Santa Cruz Counties
Cooperative Extension

Michael W. Stimmann, Statewide Pesticide Coordinator, Environmental Toxicology, UC Davis

Laura Tourte, Postgraduate Researcher, UC Davis

Louis Valenzuela, Farm Advisor, Santa Barbara County Cooperative Extension

Suzanne Vaupel, Attorney, Agricultural Economist, Sacramento

Garth Veerkamp, Farm Advisor, Placer-Nevada Counties Cooperative Extension

David Visher, Program Representaive, Small Farm Center, UC Davis

Ronald E. Voss, Director, Small Farm Center, UC Davis

Paul Vossen, Farm Advisor, Sonoma County Cooperative Extension

Al Woodard, Avian Scientist, Department of Avian Sciences, UC Davis

I
REQUIREMENTS FOR SUCCESSFUL FARMING

Introduction

Successful farmers combine their knowledge and skills with good marketing and business plans to provide the nation's food. They work with and manipulate natural resources to maximize production. Farming takes hard work, planning, discipline, prudent management, more planning, and more discipline. If you become a farmer, you must be prepared to work many hours to raise and sell your produce or products. Income will vary from year to year, depending on weather conditions, diseases and pests, and market conditions.

Most successful farmers are innovative and flexible. They acquire the necessary equipment, use appropriate cultural practices, follow the market closely, watch for new trends, understand business practices, and are astute entrepreneurs. They know that quality starts the moment they begin planning for the next year's activities.

Some people start farming part time. Some keep full-time jobs and develop their farms as part-time businesses while getting started. If you begin farming full time, you will need sufficient funds to last 12 to 18 months until you start selling your crops or livestock.

Chapter authors: Daniel W. Block, Professor, California State Polytechnic University, San Luis Obispo; William Burrows, Instructor, Shasta Community College; Shirley Humphrey, Staff Research Associate, Small Farm Center; Garth Veerkamp, Farm Advisor, Placer-Nevada Counties Cooperative Extension; Ronald E. Voss, Director, Small Farm Center, UC Davis

Commitment to Farming

Before you begin farming, carefully evaluate your reasons for farming and, if your farm will be in a remote area, consider the advantages and disadvantages of rural versus urban living. However, because of modern transportation and communications technology, rural living is less arduous than in the past. Your reasons for farming may be financial, personal, or both. You may:

- Wish to be an entrepreneur in full control of your enterprise.

- Come from a farming background or have inherited a farm.

- Like farming and want to teach your children to grow plants or raise animals.

- Wish to get away from city living.

- Wish to take advantage of investment incentives in the tax laws.

- Want to increase your income.

Running your own farm is hard work. Most farmers work more than 50 hours a week, some as many as 65 to 70. But, being your own boss allows you to be in charge of your schedule and to incorporate your family into your job.

Why I Am a Farmer

"We have a number of relatives who are farmers and I used to work on my uncle's farm as a youth. It wasn't until we were on our honeymoon when we visited one of my wife's relatives in the Midwest that I first got the 'bug.' The quality of life of this family had impressed me. We agreed that this was the kind of life we wanted to create for ourselves and the family we planned to have. The single most important factor in the quality of life I was seeking was a closeness with my family. In construction I had to be constantly on the run and away from my family. On the farm, I work where I live and can be with my wife and kids." (Mike Adams, farmer, Lake County)

"I retired after 23 years of running a produce house in Los Angeles and wanted to stay active. I turned to my 135-acre apple orchard and decided to emphasize direct marketing. I now sell in farmers' markets and distribute apple juice to stores." (Frank Rivers, apple grower, Yucaipa)

"I just finished the evening milking. This is a special time of day, the work mostly done. I relax a little and let the sights, sounds, and smells—the life of the barnyard—sing me their evening hymn, the animals eating, and the rhythm and smells of the milking, the crescent moon through the clouds, the breeze in the cottonwood trees down by the creek. They all blend to a feeling that reminds me and assures me of why I'm a farmer." (Kit Jones, vegetable grower, Lakeport)

From *Farm Link*, Community Alliance with Family Farmers, Davis, CA

The Dumonts—Getting Started

by David Visher,
Small Farm Center

Leo and Holly Dumont have a very small farm in San Martin, Santa Clara County. Craig Kolodge is their county farm advisor. He is working closely with them as they fit the pieces of the farm puzzle together. Holly says:

Without help from farm advisors like Craig, small new farmers wouldn't be able to stay in business very long. The farm advisors keep us from making terrible blunders and lead us to success.

We were going to buy a farm in Washington but my Mom had this farm, and she told us, "Come farm this one instead!" We promised we would farm for 10 years. We are three years into the project.

We have about 13 acres including pasture. We will farm about one acre this year because there are only two of us. We believe that it would be better for the economy around here and for ourselves to hire labor instead of buying equipment.

Meat production is a very important part of the farm. We might starve if we didn't have meat. In addition to keeping the areas that we are not farming weeded, they provide us with meat and manure.

We grow tomatoes, potatoes, corn, peppers, beans, 10 or 15 varieties of squash, lettuce, carrots, herbs, and flowers both for sale and to encourage beneficial insects. In the orchard we have a little of everything—apples, pears, peaches, nectarines, Asian pears, and apricots.

During the first two years we set up test plots to see what we could grow. We grew everything we could get seeds for. We thought that before we put in a 50- by 50-foot square we should see what works. It was about that time that we met Craig.

Craig was the first person to really listen to our concerns and to come up with answers. He hunts through books we can't afford and through sources we can't get to, and gets back to us. If he doesn't know, he tells us.

We know now what are the right varieties, but it's going to take more time before we become productive. This farm is an experi-

ment, but one that is so diversified it can survive the period of learning.

We have to eat what we grow here which is the other reason we need the diversity—you can get very tired of a diet of broccoli!

Like most small farmers, we need income from off the farm. I do a little writing. I teach heirloom doll making. We do house cleaning and painting when we get desperate. Any work that comes up in the off-season, we don't turn down!

If it weren't for the farmers' markets, I think farms like ours would be dead. Restaurants and other markets are fine but they want a certain thing and an exact amount and they want it now. It is hard to meet that requirement. That's why the farmers' market is great. You can sell 1,000 pounds or 10 pounds of peaches if you work at it. You don't have to make any promises that you may not be able to keep.

We sell directly off the farm on a subscription basis. We have three families that pay in the beginning of the season.

Craig Kolodge said of the Dumonts and others like them, "I see all of this land that is being broken up into small farms as an opportunity to develop a different way to look at intensively managed small farms. It is the best ground for experimentation and development of new ideas.

"Conventional wisdom works for conventional large scale farming, but it doesn't work for this kind of farming. The new models will come from small farmers themselves. Those that survive and succeed will be our teachers."

Leo and Holly Dumont with farm advisor Craig Kolodge

Sorting out personal and family goals in advance will help avoid confusion and conflict. Your goals and those of family members are important because your decisions affect every member of the family. The worksheets at the end of the book will help you evaluate your interest in farming, willingness to take risks, and farming preferences.

Farm Apprenticeships

California Certified Organic Farmers
P.O. Box 8136
Santa Cruz, CA 95061
(408) 423-2263
has a list of farms offering apprenticeships.

Center for Rural Affairs
P.O. Box 406
Walthill, NE 68067
links new farmers with older ones who are ready to retire but want to keep their farms going.

Agroecology Program
University of California
Santa Cruz, CA 95064
(408) 429-2321
offers a six-month residential apprenticeship in ecological horticulture of vegetables, herbs, flowers and fruits, including soil preparation, composting, planting, cultivation, propagation, irrigation, pest and disease identification and control.

Student Experimental Farm
Agronomy and Range Science Dept.
University of California, Davis, CA 95616
(916) 752-7645
offers a three-month, full-time summer course on organic farming and gardening, pest control, fertility, composting, cover crops, specialty crops and marketing.

If you plan to live on your farm, you need to consider such basics as availability and repair of plumbing in your house and farm buildings; cost of electricity or propane; telephone service; water availability and quality; quality of and distance to hospitals, schools, stores; road quality on and to the farm; neighbors and friends, etc. When living on a farm in a rural area, you will need to be better organized so you don't waste time and gasoline running errands. Your costs for amenities such as food, fuel, and other consumer goods may be higher in country stores than in urban or suburban chain stores. There are many books about country living in libraries and bookstores.

Knowledge and Skills

Farming involves more than planting, cultivating, and harvesting crops. It requires record keeping, financial forecasting, marketing decisions, dealing with the farm credit system, and technical knowledge about equipment, plant and animal growth, chemical use, and soils. You can acquire knowledge through written materials, courses, talking with experts and farmers, and experience. Written material is available at libraries and university and college campuses where agricultural courses are taught. Sources of expertise on farming include farm advisors, extension specialists, professors, agricultural consultants, and agribusiness representatives. Many community colleges offer short courses in small-scale and beginning farming.

Experience is a good teacher. You can experiment with a small-scale attempt to produce a crop or product, but better experience comes from working with a seasoned farmer. Many farmers offer apprenticeships that typically include room and board. A few organizations and schools also offer apprenticeships.

Skills that typify most successful farmers include a natural affinity for growing plants or raising animals, as well as mechanical, business, and human relations skills.

Finding the Right Location

Finding the right property and location should be your highest priority. Obtain information about climate, temperature, winds, soils, water quality and availability, drainage, land use laws, and the surrounding area. Has the land been surveyed and mapped? Are corners and boundaries defined? How much fencing is needed? Consider taxes, including special district and water taxes. In what condition are buildings? How much will improvements cost? Is farm labor available? How are the roads? Can your produce be transported easily?

J.R. Organics, San Diego

Is the land suitable for farming? Is the urban population encroaching and what effects will this have on your farm? This could cause problems but could also be a potential market. Carefully consider both advantages and disadvantages.

Since you will need water for crops, livestock, and domestic purposes, you should consider both the quantity and quality of water. Is it obtainable from surface or underground sources? Some areas have adequate well water. If land around you is developed, how will your water supply be affected? Contact the local irrigation district about deliveries, hook ups, and future supplies. Will you have rights to use water from streams or drill a well? Ownership of water rights is complicated. Sometimes well water can be used for irrigation, but weigh costs carefully to decide if returns justify drilling and energy investments for pumping. Are ponds an option for water storage, irrigation, livestock, wildlife, or even fish production? Are there springs? What time of year do they flow? Look again in late summer before planning to use this source of water.

Explore how waste will be disposed of. What types of waste will be farm generated? Can the wastes be disposed of on the farm? Will they be disposed of off-site? Did the previous owner have waste disposal problems? Will latent problems emerge? For example, if pesticides or fuels have been stored on the farm they may have leaked from underground fuel tanks and other containers, thereby creating polluted soil and water that is expensive to clean up. In some counties fuel tanks no longer in use must be filled or removed. Waste disposal will become increasingly important as environmental laws take effect and urban populations approach.

Chapter 2 will help you evaluate locations.

Marketing Your Goods

Before you decide which crops to grow, you need to develop a marketing plan. This should include who will buy your products and how much they are likely to purchase. You can determine this through market research. Also consider how you will deliver products. How far will you have to transport them? How much will you charge? Do you plan to sell your products through terminal markets, cooperatives, farmers' markets, on the farm, to restaurants? Answering these questions will help you decide what to produce.

Many small farmers directly market their goods at farmers' markets, roadside stands, and to restaurants, because marketing through wholesalers may eat up as much as 70 cents of each dollar consumers spend on fruits and vegetables. Direct marketing is also of great importance to small farms when their produce is too specialized, too perishable, or the volume too small to sell through conventional channels. According to one farmer, "The easiest part is to make a product that tastes good. The hardest part is distribution and sales. You have to organize demand for your product. You have to go out and sell the product before you produce it—get your markets, get your shelf space."

Read Chapters 4 and 5 for help in understanding marketing.

Keys to Success

Your economic success depends on your knowing what, how, and when to grow and market, good fiscal management, good personnel management, and a willingness to be innovative.

To be viable, a small farmer must diversify in one or more ways, such as by direct marketing; or by producing numerous crops or animals; or using an alternative production or marketing system (i.e., organic production, growing specialty crops, or adding value to your products).

By diversifying you will capture a higher percent of the price consumers pay, have year-around income, and establish a niche in the market. Diversifying may mean creating an additional on-farm business (making and selling jams, running a bed and breakfast operation, etc.) to help augment farm income.

You will have a greater chance for success if your farm's presence and its operations are acceptable to your neighbors and the community. Airplane pesticide applications, middle of the night tractor operations, the odor of manure, clouds of dust, etc. are objectionable to many people.

Faustino Muñoz, a Cooperative Extension small farm advisor in San Diego County, believes that small farmers need to maintain an active role in the community. "A number of issues, such as

The Luceros—Learning through Failure

by Pedro Ilic, Fresno County Farm Advisor

Frank Lucero looks across seven 360- by 30-foot greenhouses full of vegetable transplants and sees a source of aggravation, as well as thousands of dollars his customers owe him.

Frank and his sons, Kenny and Gary, grow grapes, vegetable crops, and vegetable transplants. But they are about to quit their transplant business after only their second season.

Frank is an old pro at farming, having successfully farmed 65 acres of grapes and vegetables for 25 years. Through careful management and long hours of work, he made a reasonable income.

Frank, Kenny, and Gary thought they were ready to tackle any challenge. Two years ago when they opened their vegetable transplant business, Kenny said, "I have been successfully producing transplants to supply my own needs for six years. I even sold a few thousand plants to growers who like their quality. I know I will be dealing in greater volume, since a lot of farmers are asking me to grow plants for them. I know we will have problems but nothing we can't solve. What can possibly go wrong?" Unfortunately, plenty!

Growing the transplants was not difficult. But, handling hundreds of growers with different vegetable crop needs, some of whom had little or no money, or spoke little or no English, was.

They found there was a vast difference among greenhouse suppliers. Some were more reliable than others in delivering on time. One supplier

Frank Lucero

took a long time delivering an automatic irrigation system and then finally shipped the wrong system.

Kenny said, "It's common with the vegetable seeds these farmers bring us to have only partially filled flats. We can't make any money this way. There is too much wasted space. We should have charged them by the flat. That way, at least our expenses would have been covered." The price they charged for transplants was too low and did not cover all costs, and, since they did not ask for a deposit in many instances, they had no protection against those who didn't pick up their plants or failed to pay.

They supplied some farmers with plastic mulch, plastic for row covers, and drip irrigation tape, at cost. This practice hurt them, because some growers didn't pay, paid late (up to seven months late), or didn't pick up their orders. Today, thousands of wire hoops for row covers are gathering dust.

Frank, Kenny and Gary Lucero are typical of many farm businesses reeling under stresses brought about by diversifying into a new, fast growing business and dealing with a large, ethnically diverse clientele, with inadequate preparation and information. The lessons they learned include:

- They didn't take enough time to find out how other vegetable transplant growers conducted business.

- They underestimated the complexities of working with a diverse clientele.

- They underestimated the amount of added labor and time the new business represented.

- They should have limited themselves to growing transplants. Other services should have been undertaken later when the transplant business was established.

- They shouldn't have accepted large orders in mid-season, which adversely impacted everybody else's orders.

- They should not they have agreed to produce transplants for which they didn't have proper facilities.

- One of them should have been assigned exclusively to deal with the clients.

protection of the environment, preservation of wildlife, land access, water availability, agricultural trade, immigration, workers' compensation insurance, and agricultural policy affect small farmers directly. The future of farming is dependent upon decisions made on these issues by policy makers and the general public, and both are increasingly removed from a direct link with agriculture. The small farmer must educate these policy makers and the public about the importance of agriculture in the community."

Muñoz suggests farmers take active roles in the local chamber of commerce, environmental groups, farmer organizations, land planning committees, community organizations, advisory committees, farmers' markets, or roadside stands.

Farming on a small scale can be enjoyable and profitable. Success depends on well thought-out decisions based on adequate information and experience. The experiences of two farm families (the Dumonts and the Luceros) provide insight into the factors that contribute to success or failure (pages 3 and 7).

Cooperative Extension County Offices

Alameda Co. Coop. Ext.
224 West Winton Ave., Room 174
Hayward, CA 94544-1298

Amador County Coop. Ext.
108 Court Street
Jackson, CA 95642

Butte Co. Coop. Ext.
2279 Del Oro Avenue, Suite B
Oroville, CA 95965

Calaveras Co. Coop. Ext.
891 Mountain Ranch Road
Government Center
San Andreas, CA 95249

Colusa Co. Coop. Ext.
P.O. Box 180
Colusa, CA 95932

Contra Costa Co. Coop Ext.
1700 Oak Park Blvd., Bldg. A-2
Pleasant Hill, CA 94523

El Dorado Co. Coop. Ext.
311 Fair Lane
Placerville, CA 95667

Fresno Co. Coop. Ext.
1720 South Maple Ave.
Fresno, CA 93702

Glenn Co. Coop. Ext.
P.O. Box 697
Orland, CA 95963

Hoopa Reservation
P.O. Box 417
Hoopa, CA 95546

Humboldt Co. Coop. Ext.
Agricultural Center Bldg.
5630 South Broadway
Eureka, CA 95501

Imperial Co. Coop. Ext.
1050 East Holton Road
Holtville, CA 92250-9615

Inyo-Mono Co. Coop. Ext.
207 West South Street
Bishop, CA 93514

Kern Co. Coop. Ext.
1031 South Mt. Vernon Ave.
Bakersfield, CA 93307

Kings Co. Coop. Ext.
680 North Campus Drive
Hanford, CA 93230

Lake Co. Coop. Ext.
Agricultural Center
883 Lakeport Boulevard
Lakeport, CA 95453

Lassen County Coop. Ext.
Memorial Building
1205 Main Street
Susanville, CA 96130

Los Angeles Co. Coop. Ext.
2615 South Grand Ave., Suite 400
Los Angeles, CA 90007

Cooperative Extension County Offices (continued)

Madera Co. Coop. Ext.
328 Madera Avenue
Madera, CA 93637

Marin Co. Coop. Ext.
1682 Novato Blvd., No. 150B
Novato, CA 94947

Mariposa Co. Coop. Ext.
5009 Fairgrounds Road
Mariposa, CA 95338-9435

Mendocino Co. Coop. Ext.
Agricultural Center/Courthouse
Ukiah, CA 95482

Merced County Coop. Ext.
2145 West Wardrobe Avenue
Merced, CA 95340

Modoc Co. Coop. Ext.
202 West 4th Street
Alturas, CA 96101

Monterey Co. Coop. Ext.
1432 Abbott
Salinas, CA 93901

Napa Co. Coop. Ext.
1436 Polk Street
Napa, CA 94559-2597

Placer/Nevada Co. Coop. Ext.
11477 "E" Avenue
Auburn, CA 95603

Plumas/Sierra Co. Coop. Ext.
208 Fairgrounds Road
Quincy, CA 95971

Riverside Co. Coop. Ext.
21150 Box Springs Road
Moreno Valley, CA 92387

Sacramento Co. Coop. Ext.
4145 Branch Center Road
Sacramento, CA 95827

San Benito Co. Coop. Ext.
649-A San Benito Street
Hollister, CA 95023

San Bernardino Co. Coop. Ext.
777 E. Rialto Avenue
San Bernardino, CA 92415-0730

San Diego Co. Coop. Ext.
5555 Overland Avenue, Bldg. 4
San Diego, CA 92123

San Francisco Co. Coop. Ext.
300 Piedmont Avenue, Bldg. C
Room 305A
San Bruno, CA 94066

San Joaquin Co. Coop. Ext.
420 South Wilson Way
Stockton, CA 95205

San Luis Obispo Co. Coop. Ext.
2156 Sierra Way, Suite C
San Luis Obispo, CA 93401

San Mateo Co. Coop. Ext.
625 Miramontes Street, Suite 200
Half Moon Bay, CA 94019

Santa Barbara Co. Coop. Ext.
105 East Anapamu, Suite 5
Santa Barbara, CA 95101

Santa Clara Co. Coop. Ext.
2175 The Alameda
San Jose, CA 95126

Santa Cruz Co. Coop. Ext.
1432 Freedom Boulevard
Watsonville, CA 95076-2796

Shasta Co. Coop. Ext.
1851 Hartnell Avenue
Redding, CA 96002-2217

Siskiyou Co. Coop. Ext.
1655 South Main Street
Yreka, CA 96097

Solano Co. Coop. Ext.
2000 West Texas Street
Fairfield, CA 94533-4498

Sonoma Co. Coop. Ext.
2604 Ventura Avenue, Room 100-P
Santa Rosa, CA 95401-2894

Stanislaus Co. Coop. Ext.
733 County Center III Court
Modesto, CA 95355

Sutter/Yuba Co. Coop. Ext.
142-A Garden Highway
Yuba City, CA 9599

Tehama County Coop. Ext.
P.O. Box 370
Red Bluff, CA 96080

Tulare Co. Coop. Ext.
Ag. Building, County Civic Center
Visalia, CA 93291

Tuolumne Co. Coop. Ext.
2 South Green Street
Sonora, CA 95370

Ventura Co. Coop. Ext.
702 County Square Drive
Ventura, CA 93003-5404

Yolo Co. Coop. Ext.
70 Cottonwood Street
Woodland, CA 95695

II
THE BASICS

Introduction

If you plan to farm, start acquiring information now. You need to know about land availability, price, and zoning; about soil quality; about water sources, quantity, quality, and cost; and about the climate. How will you dispose of wastes? What equipment will you need? How are the fences and gates? Will you need to build animal shelters? Is there a workshop or shed? In what condition are the roads—both on the farm and those you'll use for access to markets? Determine whether adequate labor is available when you'll need it. Look into availability of electricity and other sources of energy. The initial decisions you make in developing a farming enterprise will influence the ultimate success of your operation. The time, effort, and money involved in planning are well worthwhile.

Land

Obtaining a good piece of land for your agricultural operation is critical. The land must be in a climatic zone that is suitable for your planned commodities or livestock. The appropriateness of a piece of property for your operation will depend upon the slope of the land, as well as the depth and texture of the soil and the amount of water the soil holds. Other factors to consider are drainage, susceptibility

Chapter authors: Aziz Baameur, Farm Advisor, Riverside County Cooperative Extension; Steve Blank, Extension Economist, Agricultural Economics Department, UC Davis; Faustino Muñoz, Farm Advisor, San Diego County Cooperative Extension; Richard Smith, Farm Advisor, San Benito, Monterey, and Santa Cruz Counties Cooperative Extension; Paul Vossen, Farm Advisor, Sonoma County Cooperative Extension to flooding, salinity problems, and indigenous diseases. The type of land you buy affects the type of irrigation system you use. Drip irrigation has opened up sloping land to many commodities such as strawberries, vegetables, and tree crops. Furrow and sprinkler irrigation are more suited to flat land.

Finding good land is not simple. Much of the good flat land is already under production or paved over.

Land costs for either rental or purchase are high in California. Usually, land is more costly along the coastal valleys and less costly inland. The quality of the land influences the price. Sloping land or land with very heavy soils is less expensive. Inexpensive land may limit what you can grow, when you can grow it, and productivity. Because of the scarcity of prime farmland, some commodities are now grown in locations not thought suitable previously (prune growing has been moved from Santa Clara to Butte County and avocados are now grown on hillsides).

Land dry in summer can become a marsh after winter rains and spring rain runoff. If you see sedges, rushes, or other moisture-loving plants, check for high water tables or hardpan soils that restrict drainage. Ask neighbors about potential flooding and drainage problems and visit the prospective acreage at different times of the year. Ask your local US Department of Agriculture's (USDA) Soil Conservation Service (SCS) for information.

Renting or Purchasing Land

You do not need to own your farm. If you can't or don't want to buy land, you can lease private or public land, share-crop, sub-lease from another grower, or enter a joint venture. The Farmers Home Administration (FmHA) can help you purchase repossessed land at a reduced cost. See Chapter 3 for information on assistance available from the FmHA.

Zoning

The three categories of land zoning are agricultural, industrial, and residential. Usually, only land zoned "agricultural" can be used for farming. When looking for land, consider the current zoning of land surrounding a parcel. Remember, zoning may change over time. Also, if a parcel is near a residential area, restrictions may be placed on its use. The use of herbicides and pesticides are restricted by some cities for agricultural parcels adjacent to residential areas. Look for land that is not likely to be in the path of residential or industrial development during the period over which farming is planned. The high cost of developing farmland (establishing buildings, irrigation systems, and other improvements) usually requires that the parcel be in production for 10 to 20 years to be profitable.

County planning commissions establish zoning for land use. City planning commissions influence zoning in or near their boundaries. Owners can petition planning commissions to have zoning changed. In cases of annexation by a city or of eminent domain (the power of the government to take private property for public use upon compensating the owner), land can be rezoned without the consent of its owner. Owners can appeal zoning changes to the planning commission and through the courts. Historically land has changed from agricultural zoning to residential or industrial use, not the reverse.

Williamson Act

The Williamson Act (Land Conservation Act of 1965) offers property tax relief to producers who preserve agricultural land from development for a set time (usually 10 years) in return for lower property taxes. Participation is voluntary for both landowners and counties or cities. Contracts can be terminated through nonrenewal, cancellation, eminent domain, or city annexation.

Agricultural and other open space lands are eligible. Location within an agricultural preserve is required, as well as minimum parcel sizes and other conditions set by local governments.

Participating acreage is classified into three land use categories—urban prime, other prime, and nonprime:

- Prime land is classified I or II by the Soil Conservation Service; rated 80-100 in the Storie Index Rating; used for livestock with an annual carrying capacity of at least one animal unit per acre; used for trees and vines which earn, during the commercial bearing period, at least $200 per acre, or used for unprocessed agricultural products which earn at least $200 per acre for three of the previous five years.

- Urban prime land is located within three miles of cities of at least 25,000, or sometimes, 15,000-25,000.

- Non-prime land is all other agricultural land.

For more information on the Williamson Act, contact your county planning department or American Farmland Trust, (916) 753-1073.

Size

The amount of land you need to buy or rent depends upon your planned enterprises, capital resources, and anticipated profits. Some commodities, such as high-value herbs and specialty crops, give more return per acre than others, such as wheat or barley, and can be grown profitably on small acreages. Marketing is the key because the more you earn per acre, the less land you need to make a living. Your business plan, long-range plan, and need for immediate revenue from the farm will help you make a decision.

Local county farm advisors, farmers, and farm consultants can help you estimate potential yields and prices. The Federal-State Market News Service, a compilation of commodity market prices, supply, and demand, collected from major terminals and other handlers, can give you estimates of market prices and trends. Call (900) 555-7500. (There is a charge.)

Soil

Take time to consider the quality of soil you will need to obtain profitable yields.

The best fruit and vegetable soils are sandy and fairly deep with excellent drainage. The ground can be worked early in the spring and soon after an irrigation or a rain. This is critical to satisfying the niche markets small-scale operators seek because the ground can be worked earlier or later in the year or more days per year than heavier soils.

Soil tests will provide important information on soil. See Chapter 7 for more information.

Poor subsoil drainage can be a problem. Drain lines may be necessary in heavy soil for crops such as citrus. Surface drainage may also be needed. In drip or sprinkler irrigated orchards, furrows between the rows leading to ditches or drain pipes to collect surface runoff can be helpful. In severe cases, erosion control or grade stabilization structures may be necessary to collect soil and water runoff and to divert the water to drain systems.

SCS provides soil survey maps and conservation plans which show soil types and possible uses. The maps are available for most areas of California. The staff explains soil quality and type and offers suggestions for irrigation, drainage, erosion control, and tillage methods. For some conservation practices (irrigation, drainage, and erosion control) cost-sharing financial help is available from USDA's Agricultural Stabilization and Conservation Service (ASCS). Your county University of California Cooperative Extension farm advisors as well as the county's Agricultural Commissioner can also help. Look in your phone book under County Government Offices to find them. Look under US Government Offices to find SCS and ASCS. Other sources of information are local farmers, farm consultants, farm suppliers, well drillers, and equipment dealers.

Diversified cropping and landscape

Water

Water is the most critical element in farming. Without water there is no farming. Most beginning farmers are baffled by how much jargon and math go into figuring a plant's water needs. Unfortunately, science can only help you make estimates.

In planning a new operation, you'll need to know about water sources or supplies, the quality of the water, how much water is available, how much you'll need, and the possible methods of irrigation. See Chapter 7 for information about the quantity of water needed and different types of irrigation.

Water Sources

Water costs vary substantially from one location to another depending on the source of water—surface, water districts, or wells. The cost of your initial farming investment will be influenced by the water situation. Also consider whether you will need on-farm water storage. Are facilities in place or will you have to build them?

If surface water (ponds, lakes, streams, rivers) is available, find out its flow fluctuations during the year. Does it go dry during the summer? Will water be available when you need it? What are your rights to the water on the farm? The California Department of Water Resources in Sacramento has current and historic data on river and stream flow.

If there is no surface water, you may have a water supplier (irrigation or water district). Access to California's intricate statewide dam-reservoir-canal systems is controlled by irrigation or water districts. The farmer-users pay for their water with district fees. To locate the district in your area, ask other farmers, your farm advisor, or the local Farm Bureau. It is a tremendous benefit if land is located within a district, not only for access to water, but for reducing capital outlay for irrigation equipment.

Perhaps you have a well or will have one dug. Check into legal rights to underground water. Ask the well driller about drilling depth, amount of water expected, cost of drilling, cost of pumping at given depths, and well design. Other considerations include pressure, meter capacity, reliability (is water available year around?), and volume.

Regardless of the source of water that is available, you will need to investigate the costs of the water and add that information to your production budget.

Water Quality

Water quality is a measure of elements in the water. Salts are the greatest concern in water quality, because high salt levels can harm plant performance. Major culprits are chlorine, sodium, calcium, and nitrates. To determine the quality of water, have it analyzed by a laboratory. A list of laboratories is in the publication California Commercial Laboratories Providing Agricultural Testing (Pub. 3024), published in 1991 by University of California ANR Publications, 24 p.

Salt is measured in parts per million (ppm) and by the electrical conductivity (EC) of the water. Crop sensitivity to salts varies tremendously. With drip irrigation, it may be possible to use water with 1,000 to 1,500 ppm total salt. With sprinkler irrigation, 800 to 1,000 ppm is the maximum range. Salts in solution conduct electricity. Electrical conductivity readings of your water supply are an indicator of water quality and can be converted into parts per million. An EC of 1.0 is approximately 680 ppm salts. Generally, water with an EC below 1.0 is preferred. Concern should be noted for boron levels above 0.5 mg per liter, salt levels above an EC of 0.75 and sodium and chloride levels more than 3.0 mg per liter. Other contaminants to consider, depending on the location of water sources, are sand, algae, and industrial wastes. These can be harmful to the plants or the irrigation system.

Climate

Climate will strongly influence your selection of an enterprise. Many plants have specific temperature, day length, and developmental limitations that cannot be exceeded. Some plants cannot tolerate freezing temperatures and others will not set or retain a crop if temperatures get too high. Plants that are sensitive to extreme winter cold are called "tender" or "not winter hardy."

Major climatic influences vary considerably and are determined by elevation, marine or fog influence, wind patterns, rainfall, slope or exposure, frost-free days, average temperatures, and temperate extremes. Investigate all of these factors.

You can get historic and current weather data through the Department of Water Resources CIMIS program and the University of California's IMPACT system.

CIMIS (California Irrigation Management Information System) is managed by the California Department of Water Resources. There are automated weather stations located throughout the state that gather information on rainfall, wind direction, temperature, evapotranspiration, humidity, etc., to help growers plan irrigation. The information is fed into the main computer system in Sacramento hourly. CIMIS can be accessed through several computer networks, or you can have an individual account. To use the system, you need a computer, communication software, and a modem. System users are assigned passwords and identification numbers and receive literature and a map showing where the stations are located. Users can receive advice over the telephone. For information, or to apply for an account, call the Department of Water Resources at (916) 327-1836.

Cool-Season Vegetables

Hardy	Half-hardy
Asparagus	Beet
Broad beans	Carrot
Broccoli	Cauliflower
Brussels sprouts	Celery
Cabbage	Chard
Chives	Chicory
Collards	Chinese cabbage
Garlic	Globe artichoke
Horseradish	Endive
Kale	Lettuce
Kohlrabi	Parsnip
Leek	Potato
Mustard	Salsify
Onion	
Parsley	
Peas	
Radish	
Rhubarb	
Spinach	
Turnip	

Warm-Season Vegetables

Tender	Very Tender
Cowpea	Cucumber
New Zealand spinach	Eggplant
	Lima bean
Snap bean	Muskmelon
Soybean	Okra
Sweet corn	Pepper, hot
Tomato	Pepper, sweet
	Pumpkin
	Squash
	Sweet potato
	Watermelon

From Lorenz, Oscar A. and Donald N. Maynard. 1980. *Knott's Handbook for Vegetable Growers* (3rd ed.). New York, NY: John Wiley & Sons, Inc. 390 p.

The University of California's IMPACT system provides growers information on pest management and on daily weather information from CIMIS as well as other data. To apply for an IMPACT account, call University of California Integrated Pest Management (UC IPM) at (916) 752-8350. To use the system, you need a computer, communication software, and a modem. The account and the software are free. The instruction manual costs $15, and, of course, all phone charges are the responsibility of the customer. If you do not want a personal account, you can get information from your local Cooperative Extension office, which is connected to the IMPACT system.

Vegetables

Cool-season vegetables grow best in areas or times of year where temperatures don't exceed 75° to 80° F. Warm-season vegetables require soil temperatures above 50° F to grow properly and long, hot growing seasons. Cool-season crops typically are frost tolerant. They germinate at cooler temperatures, have shallower root systems, and have smaller plants than warm-season crops.

Livestock

Livestock and poultry producers also must consider weather. For example, poultry producers must provide a heated facility for young birds during cool weather. As the birds mature, they may become overheated in warm climates, in which case the facility must provide cooling also.

Fruits

With fruit crops, chilling (the number of hours below 45° F) is an important climatic factor that influences bud break, fruit set, and fruit development. Most fruit varieties require from 200 to 2,000 chilling hours in the winter to break dormancy. Some early blooming fruit trees are more susceptible to late spring frosts than to very cold January or February temperatures. Summer high temperatures can be deadly to crops that have limited heat tolerance during certain stages of development, such as raspberries.

Apricots were historically grown in the Winters area of northern California because of the moderate climate that was influenced by cool breezes from the San Francisco Bay area. More recently, with variety changes, the Fresno area has captured the early and best market for apricots. Berries (strawberry, raspberry, blueberry, and blackberry) perform better under cool coastal climates. Wine grapes, however, will not mature if the climate is too cool, yet, if too warm, quality is compromised.

Citrus requires a great deal of summer heat and relatively frost-free sites, so it usually is grown on slopes just above valley floors in the southern part of the state. To grow citrus, choose a location where damage from frost is infrequent and light to avoid fruit losses. Otherwise you may need heating or wind machines. The cost of this equipment should be considered and added to the budget. Fuel and power cost increases are making this a more critical consideration. You may wish to record temperatures with thermographs or thermometers through one or more seasons before starting your farming venture to get an idea of frost conditions.

Waste

As a farmer, your role as a steward of the land becomes more apparent. Your decisions on handling waste directly affect the environment. Re-using and recycling reduce the amount of waste.

Locate the nearest landfill for disposal of large items and non-recyclable items. Investigate local burning regulations. Can you burn? What can you burn? What times of year is burning likely to be restricted? Does the house have a septic system? Additional information is available from your county health department, SCS, and Cooperative Extension offices.

Disposing of hazardous wastes is a continuing concern for farmers. Even if you plan to use no chemicals for fertilizing or pest control, there may be partially used or unopened containers left by previous occupants. Disposal is strictly regulated to protect the water and air supply. Ask your county Agricultural Commissioner, Cooperative Extension farm advisor, or the Farm Bureau about disposal sites and dates.

If you plan to raise livestock, your operation will need a system for waste disposal. If you plan to incorporate manure into the soil, study composting and proper application to avoid burning plants or overloading the soil with nitrogen. Will you sell or trade excess manure to neighboring farmers?

Antonio LaScala, Sonoma County farmer

Equipment

Appropriate equipment is essential to farming. It is also expensive to buy and maintain. Many growers borrow equipment back and forth or hire people who own specialized or expensive equipment. Renting equipment is another option. Buying used equipment will help, but be aware of the disadvantages as well as the advantages. Include your equipment needs in your farm plan. Make a list of essential items and get them first. Acquire other items as your budget allows or as you find good deals.

New Equipment vs. Used Equipment
New equipment is reliable and it has a warranty. It also costs more than used equipment. Some growers with limited mechanical skills buy basic equipment new and purchase less complicated equipment used. Mechanically skilled growers often buy used basic equipment.

Used equipment is less expensive than new equipment; however, the equipment may have a tendency to break down. The more mechanical skills you have, the more money you can save. Community colleges offer classes in farm machinery, welding, and equipment repair. Welding classes are especially useful for learning to repair, fabricate, or modify equipment.

Find used equipment by looking through classified ads in local newspapers and farm magazines and talking with grower friends. Local tractor dealers often have reconditioned used equipment. Auctions are important outlets for used equipment. Talking about your equipment needs at farm meetings often results in useful leads.

Depreciation Schedule				
Machine	Current Price	Useful Life (Years)	Investment Per Acre Acre(40-acre farm)	Depreciation Per Acre
Tractor, 70 hp	$ 15,000	10	375	38
Crawler, 50 hp	50,000	15	1,250	83
Irrigation system	18,000	15	450	30
Plow, 4-16	2,000	10	50	5
Cultivator, 4-row	2,600	10	65	7
Planter, 2-row	1,500	10	38	4
Tandem disk, 10'	4,600	10	115	11
Pickup	9,000	10	225	45
Total	102,700	5	2,568	223

From: Takele, Etaferau, *How to Determine Your Cost of Production*, Pub. 011, ANR Publications.

Equipment Resources and References

Fundamentals of Machine Operation Series: Planting; Tractors; Tillage; Machinery Management. 1981-1987. Deere and Company, Customer Service, Distribution Service Center, 1400 13th Street, East Moline, IL 61244; (800) 522-7448.

Hot Line Farm Equipment Quick Reference Guide. Heartland Communications Group, Inc., l003 Central Avenue, P.O. Box l052, Fort Dodge, IA 5050l; (800) 247-2000. For farm tractors and combines only; monthly issues cover other farm equipment values.

Ten Tips for Spotting Cover-ups when Buying Used Equipment

By James W. Garthe,
Department of Agricultural Engineering, Pennsylvania State University

Most sellers are reputable and forthright. But you should be aware of gimmicks, tricks, or deceptive selling practices. Cover-ups are often difficult to distinguish from oversights or sloppy maintenance practices. But, regardless of intent, decide whether you want to buy used equipment about which you cannot differentiate between poor maintenance and questionable selling practices.

1 Fresh paint gussies it up. A new coat of paint does wonders to make equipment look good. Don't avoid all repainted machinery, but study what is before you. Painted over safety or instructional decals should make you wonder if this lack of carefulness typifies the history of the machine. Although sandblasting is perhaps the best way to prepare metal for painting, the high speed sand grains can damage grease seals and bearings. A properly sandblasted machine should be disassembled, blasted, and cleaned before reassembly and painting. Attention to detail in painting is a sign that the seller feels the machine merits this outlay of time and effort.

2 Glitz! Look for new parts, especially seats, starters, batteries, steering wheels. Ask what happened to the old parts. Perhaps they were weather-beaten, destroyed in an accident, or worn out. What else was damaged along with the replaced part? Ask questions. Ask for maintenance records.

3 Is the product identification number (PIN) visible? These tags, also called end product serial numbers, are placed on the machine by the manufacturer and registered to the owner. Tampering with these numbers is illegal. Sometimes these plates or decals are missing, mutilated, or painted over. If the owner cannot prove ownership, look elsewhere.

4 Too clean? There's nothing wrong with a clean machine unless tell-tale indicators of wear are washed off. Gaskets or oil and grease seals that leak badly can be cleaned. The paint may be like new in these areas since it was protected by years of oil and dirt build-up.

5 How do the components fit together? Sometimes components are assembled hastily. Close inspection will reveal hammer marks, sloppily fitting bearings or bushings, misaligned sheaves, twisted or kinked hoses, and so on. Ask to see the owner's and parts manuals to be sure the right part is in the right place.

6 Watch for tricks with oil. Sometimes oils have been added or switched among machines, and unfortunately there is no easy way to detect this. In some cases, thicker oils are added to crankcases, gear cases or hydraulic reservoirs in an effort to reduce leakage. Look at the oil, feel it, smell it. Additives often have a tackiness agent which can be detected by tapping the oil between your fingers to see if it clings.

7 "It was just overhauled!" Ask for details on exactly what was overhauled. Shop receipts are invaluable. On tractors, look for edges of new gaskets or shiny metal at parting lines of assemblies. Make sure all areas of disassembly and assembly were cleaned thoroughly.

8 Is the salesperson reputable? A retail dealer may have a tractor repaired to re-sell. If the repairs are not performed during slack winter months, it's a signal the dealer isn't convinced the repairs are cost effective. Take your time when considering such machinery. The salesperson should allow you time to decide whether to buy his used unit.

9 Was it used for contractors' equipment? Agricultural tractors are often bought for industrial or construction use because they cost less than heavy duty industrial equipment. Transmissions, frames, front axles and bearings, brakes and hydraulic systems are far more severely strained in this type of service than in farm use. Look for signs like oversized or extra-duty (8- or 10-ply) front tires. Look for backhoe or fork-lift attachment pads or wear on the frame. Maybe you'll detect yellow or orange industrial paint underneath a topcoat or in an obscure place.

10 Run it and drive it. If you're buying a tractor, run or drive the machine. If the seller leaves to go get the tractor, go with him. The start-up of a machine is an important clue to the condition of the engine, just as operating under load is a good indication of engine and drive train performance. On a diesel, if the glow plug or starting fluid are needed for a mild day, there will be complications. Pay attention to sounds as the engine warms to operating temperature, then load it down in all gears. Have a friend drive the machine so you can watch and listen to it.

Use these tips together with patience, perseverance and, most of all, inquisitiveness. Don't be afraid to ask for owner's manuals, service manuals and maintenance records.

From: *Farm Machinery Days for Small and Part-time Farmers,* Pub. 45, Northeast Regional Agricultural Engineering Service, Ithaca, New York.

I and T (Implementation and Tractor) Farm Equipment Trade-in-Guide and Farm Tractor Trade-in-Guide. Technical Publications Division, Intertec Publishing Corp., P.O. Box 12901, Overland Park, KS 66212.

Sample Set of Vegetable Production Equipment

- Bed shaper
- Lister
- Seeder (many types)
- Cultivating sled
- Lilliston
- Transplanters (many types)
- Disc
- Harrow (bed)
- Chisler
- Sprayer
- Tractor (wheel tractor, 60-70 horsepower, four-wheel drive if you can afford it). A small disc can be pulled with this tractor but a larger disc should be pulled with a 90-100 horsepower tractor, or rent a larger tractor.

Buildings and Roads

When you evaluate a farm for purchase or to rent, consider how much you will have to spend to make buildings and roads fit your operation. A catalog listing USDA plans for farm buildings and other farm structures is available for reference in county Cooperative Extension offices. Plans may be ordered for $4 a sheet from Plan Services, Agricultural Engineering Extension, University of California, Davis, CA 95616; (916) 752-0120.

Will area roads be passable at all seasons? Will you need to build roads on your property? A 70-page Road Building Guide is available for a fee from Mendocino County Resource Conservation District, 405 Orchard Avenue, Ukiah, CA 95482.

Labor

Farming involves hard work and people to do it. Assess the labor supply available—yourself, your family, hired and bartered. A business plan will help you determine what type of enterprise to develop considering the labor supply available. It takes time, skill, and resources to develop a competent labor force capable of consistently performing quality work.

Some crops, such as vegetables, herbs, and strawberries, are labor intensive throughout the season. A vegetable farm could require 12 to 15 full-time workers, while an apple orchard might need only two people except for two weeks of harvest. Request production cost sheets from your local Cooperative Extension farm advisor's office to determine labor needs and costs of growing a specific crop in a particular area.

Managing labor and maintaining records are critical to farming. Be aware of documentation relating to nationality, insurance, benefits, and labor laws. Careful record keeping can help you train and evaluate workers, keep them safe, and observe their legal rights. You may also need to plan for housing and medical needs. Many farmers plan crop cycles so they can keep a central, steady core of skilled workers busy throughout the year.

More ideas for successful labor management are in Chapter 11.

Energy

Don't forget to explore energy access and cost. You can contact the local utility company to find out how much electricity and gas in your area cost. They may be able to give you information on past use at your location.

Your farming profit, the difference between input costs and selling price, can be increased with carefully planned energy use. Energy is used in almost all aspects of farming, although different enterprises require different amounts and kinds of energy. Your equipment uses energy. All forms of irrigation use energy. Pesticides and fertilizers use energy while being produced—you pay for this energy in your purchase price. Postharvest storage and transportation of your goods to market use energy.

Careful planning of energy and irrigation systems, conservation, reduction of chemical inputs, integrated pest management, minimum tillage, and improvement of tilth will save you money. The California Energy Commission sponsors a farm energy assistance program to help combat surface water shortages and groundwater contamination, and to encourage reduction of electricity and petrochemical use. They offer low-interest loans for purchase of energy efficient farm machinery. For information, write: 1516 Ninth Street, Sacramento, CA 95814-5512.

Additional Resources and References

Agricultural Resources of California Counties. 1982. Pub. 3275. Oakland, CA: ANR Publications. 134 p. Shows historical development of agriculture by county, average farm size, average farm income, overall economic conditions, etc.

Farm Management. Pub. ANRP 011. Oakland, CA: ANR Publications. 26 p. Four booklets: Considerations in Enterprise Selection, Farm Leases and Rents, How to Determine Your Cost of Production, and How to Finance a Small Farm.

Generalized Plant Climate Map of California. 1988. Pub. 3328. Oakland, CA: ANR Publications.

III
FINANCES

A farm is like any other business: it must generate a profit or it will fail. The best insurance against failure is to plan in advance and keep accurate, up-to-date records.

Financial Records

Your farm records are important farm tools that, when diligently kept, will provide information needed to make decisions to improve the performance of your business.

You need records to show increases and decreases in assets, liabilities, revenue and expenses. You need inventory records, depreciation records, records of loans and mortgages, payroll records, etc. Forms and record books are available in stationery stores, farm management texts, accounting textbooks and magazine articles. Lending institutions often have their own forms.

With completed records, you will have a great deal of information. The records will help you determine the profitability of the farm, provide information to use when applying for loans, suggest ways to increase earnings, provide data for taxes and other reports, provide information on crop yields, and so on. Your year-end analysis

Chapter authors: Karen Klonsky, Extension Specialist, Agricultural Economics Department, UC Davis; Steven Blank, Extension Economist, Agricultural Economics Department, UC Davis

Net Worth Statement

Current Assets	Current Year Book Value	Current Year Market Value
Cash on hand and in bank	5,870	5,870
Savings accounts	5,800	5,800
Stocks, bonds, securities, mutual funds	0	0
Accounts receivable	8,100	8,100
Inventory, products, supplies	101,279	101,279
Inventory, growing crops	10,000	10,000
Miscellaneous assets	36,000	36,000
Total current assets	**167,049**	**167,049**
Long term assets		
Land	1,250,000	1,250,000
Trees and permanent crops	0	0
Buildings, improvements	259,480	259,480
Farm equipment	33,737	45,000
Tractors, trucks, autos	71,221	75,000
Breeding stock purchased	2,361	2,361
Inventory, livestock	624,900	624,900
Personal property, household goods	32,200	32,200
Real property, off-farm	0	0
Other misc. assets	12,000	12,000
Total long-term assets	**2,285,899**	**2,391,461**
Total assets	**2,452,948**	**2,558,510**
Current liabilities	**435**	**435**
Current portion of long-term debt (farm)	57,000	57,000
Current accounts payable (non-farm)	200	200
Current portion of long term debt (non-farm)	0	0
Income tax liability	18,000	18,000
Total current liabilities	**75,635**	**75,635**
Long-term liabilities	**0**	**0**
Accounts payable less current farm	0	0
Accounts payable less current non-farm	0	0
Long term debt less current farm	418,000	418,000
Long term debt less current non-farm	0	0
Total long-term liabilities	**418,000**	**418,000**
Total liabilities	**493,635**	**493,635**
Net worth	**1,959,313**	**2,064,875**
Total net worth and liabilities	**2,452,948**	**2,558,510**

will show whether your success or failure was a result of factors you could or could not have controlled. The results may be due to management decisions, economic conditions, or farm resources. This knowledge will help you decide what can be done to improve your operation in the future.

One of the fundamental instruments that will be developed from your farm records is the net worth statement. The net worth statement is a listing of the assets and liabilities of the business. Liabilities are the liens on the property, or the amount of money owed to others. Assets are the capital owned by the business that have market value. They can be valued by their purchase price in the year of acquisition minus the depreciation to date (book value) or the price the asset would receive if sold at the present time (current year market value).

Changes in net worth show the growth or downsizing of the business. Changes in net worth using market value also include changes due to inflation.

Your Business Plan

Developing a business plan will force you to think through what you want to undertake. What are you going to produce? Where and how are you going to sell it? Who will be responsible for the different jobs that have to be done? What is your financial situation?

The time invested in developing a business plan is worthwhile because 1) it makes you more aware of the strengths and weaknesses of your business, and 2) it significantly raises the chances that lenders/investors will fund your proposal.

A successful business plan is logically organized. Include enough information to convey your message, but not so much that you overwhelm and confuse your readers. Lenders, investors, and farm managers need facts and figures to help them decide whether to fund a new or ongoing operation or project. The data is usually presented as a one-year, three-year, or five-year business plan.

The most effective business plan will have a business charter, an explanation of management structure, product information, a market plan (see Chapter 4 also), and a financial plan.

Business Charter

A business charter is a set of statements defining who, what, when, where, and why about your business:

Who you are.

What your purpose is; what your products or services are; what markets you will use; what market share you expect to capture; what profits you expect.

When you are going to start operating; when you expect to achieve your goals.

Where you are going to sell your product or service (locally, regionally, etc.).

Why your business is being formed.

In this sample charter for a nursery, all five "w's" are included. The statements are simple and specific.

Management Structure

The management structure indicates who will be responsible for production/ farming operations, marketing/sales, and finance/administration. A flow chart or

---

Business Charter: Green Thumb Nursery (who)

Green Thumb will engage in wholesale and retail nursery sales in Mendocino County and surrounding markets (what/where).

John and Jane Doe will be partners and sole owners of Green Thumb (who) .

Green Thumb will specialize in plants especially suited to the climate of coastal counties, carrying a wide assortment as required by market demand (what/why).

We will achieve a 10 percent share of the retail nursery market in Mendocino County within four years and 15 percent during the fifth year (what/when/where).

We will achieve a 39 percent profit margin on retail sales during the first quarter of the first year of operation (what/when).

We will earn pretax profits of 18 percent on sales beginning the fourth quarter of year two and continue through the fifth year at this level (what/when).

We will provide a positive cash flow beginning the first quarter of year three (what/when).

We will produce a net return on investment of 250 percent by the end of year five (what/when)

We will expand the wholesale produce and landscaping service line as market demand and profits permit (what/when).

diagram showing lines of responsibility may be used to illustrate the chain of command. If one person will have responsibility for more than one management area (as is typical on farms), estimates of what percentage of time will be spent in each area are useful to both lenders and managers. These estimates may show that insufficient time is available for a person to adequately fulfill some responsibilities, and the plan will need to be revised. Lenders want to know that workers understand the scope of each position, so include short job descriptions. (See the management structure worksheet at the end of the book.)

Publications on Farm Financial Management

California Farm Record Book. 1983. Pub. 3342. Oakland, CA: ANR Publications. 76 p. Instructions and record space for farm income and expenditures.

Kamoroff, Bernard. 1990. Small-Time Operator. Laytonville, CA: Bill Springs Publishing. 190 p. Worksheets and a guide to how to start your own business, keep your books, pay your taxes and "stay out of trouble."

Management is the most significant profit-determining factor. You must have competent management to plan, organize, direct, staff, and control your operation in order to make a profit. Lenders will carefully study the management structure before funding a business. Sometimes a lender sees that a business could succeed if the owners were not the managers—something the owners are unlikely to realize.

Product Information

A business plan includes a product description. Keep industry jargon to a minimum. For potential lenders or investors, only a general understanding of most agricultural products is necessary. Some lenders may want a brief description of the inputs (such as land and water) needed and the production processes.

Marketing Plan

Describing the market for a product or service is difficult but necessary. The lender must have it to evaluate your proposal. A detailed marketing plan is particularly important for a new or unusual commodity. It shows you've done your homework and understand the potential for your enterprise. Be conservative in your planning so there is a margin for unforeseen setbacks.

To analyze the market potential for a product, determine the current and potential consumption of the product or service, the types of markets to be used, the types of distribution systems to be used, how the market is entered, the types of buyers that must be satisfied, the types of selling arrangements you will use, and the prices you will charge.

See Chapters 4 and 5 for more information on marketing your products.

Financial Plan

Financial records and reports are the key to a business plan. They show whether you should proceed, stop and reevaluate the situation, or cease operations altogether. The financial records a lender will probably ask for include an income statement, a cash flow statement, depreciation schedules, income tax information, and a balance sheet. Sometimes other financial reports are requested.

When you prepare these reports, you will be determining your investment capital needs (the amount of money you need to pay for facilities and equipment) and your

FINANCES

27

working capital needs (the amount of money you need to run your business). From this information you can decide how much, if any, money you need to borrow.

Borrowing Money

The cost of borrowed capital is a significant part of most farmers' budgets. You should look for the best price for loans just as you do for fertilizer, seed, and other expenses. Banks compete for good loans by offering lower interest rates, lower loan fees, or other services. Compare loans carefully.

Nothing is more important when working with a lender than being prepared. Don't ask, "How much can I borrow?" This shows you do not have a budget or a clear understanding of your cash needs. Your financial plan will clarify your borrowing needs and repayment ability. The lender may have a list of the financial documents you need to provide. You may be asked to fill out financial questionnaires. If you don't have some financial information available, you must prepare "pro forma" statements showing the expected results.

Documents the Lender May Request

■ **Resume** showing your agricultural education and experience.

■ **Business plan.**

■ **Financial records** for each of the last three to five years, including:
Balance sheets listing assets and liabilities.

Income statements (profit and loss statement).

Cash flow statements showing when and how much money is required for day-to-day operations, and when and from where it comes. You may need a loan if income lags behind sales by 30, 60, or 90 days—the customers' payment period.

Income tax statements (individual, partnership, corporation).

Collectible notes and accounts receivable (dates, sources, terms).

Information on outstanding loans (lender, terms, conditions, account numbers, interest rate, maturity).

■ **Property information**
Equipment serial numbers and descriptions.

Titles to real estate and personal property (such as machinery), particularly if these assets are to be used as securities for the loan.

Insurance policies for equipment, liabilities, and crops (carrier, policy number, amount of coverage).

■ **Farm business information**
Value and quantity of crops listed as assets (estimated by outside party).

Cropping plans and map of fields (including water sources, location of wells, pump information, planting dates, varieties for permanent plantings).

Soil maps.

Lease agreements (including the cash rent or share crop agreement and what will be grown on leased land).

Marketing plans and contracts (including sales contracts and cooperative memberships).

Three to five years of production history.

Questions to Ask a Lender

1.   Can real estate be used as security? What is the appraisal fee?

2.   Is there an application or commitment fee? Is the fee returned if the loan is made? Is it returned if the loan is not made?

3.   Are there closing costs, inspection fees, charges for documents and who pays for them (usually paid for by borrower)? Can they be made part of the loan?

4.   Will maintaining a bank account at the lending institution reduce the cost of the loan?

5.   What are the rate/term options?

Questions a Lender May Ask

1.   Are you a co-signer, endorser, or guarantor for debts incurred by others?

2.   Do you owe income tax?

3.   Are you involved in pending lawsuits?

4.   Are you involved in any contracts? What happens if you fall short of your contract?

5.   Do you have an established market for your products?

6.   What are the long-range plans for your operation, including changes in crops, size of operation or ownership.

7.   Is there potential liability in connection with violation of local, state or federal laws or regulations?

8.   Most lenders ask you to fill out an environmental questionnaire asking whether there have been pollution problems on the property, the history of chemical usage on the property, whether there are underground fuel tanks or chemical storage areas on the property.

9.   What are your sources of water? Cost of water?

10.   What risk management techniques are you using with respect to production, marketing, and finance?

There is no substitute for a good working relationship with your lender. Creative financing such as restructuring a loan, or refinancing or rolling over production loans are done at the discretion of the lender. A long-term relationship with a lender increases your financing options.

Sources of Credit and Information

If you are a beginning farmer, you may be having trouble getting a loan. Perhaps you do not have enough money or assets to use as loan security. Maybe your bank does not make small loans because the costs are as high as for large loans, but the profit, based on the amount lent, is smaller. Other problems include the high risk of growing specialty crops.

Despite these problems, credit is available from banks, the Farm Credit System (FCS), Farmers Home Administration (FmHA), and rural development corporations. These groups coordinate their efforts to develop loans that will appeal to small farmers as well as lenders.

Before applying for a loan, arrange for a co-signer (if necessary) who has the required collateral, and ready your business plan for the lender.

Local Banks

Local banks are a source of short-term credit. Because the primary source of funds is from deposits, loan size may be small. Some local banks are able to offer long-term loans by selling the loans to an insurance company that provides the funds for the loan. Loan fees and the interest rate may be reduced if you deposit your profits back in the bank.

Farm Credit System

The FCS offers credit-related services to farmers and ranchers through three associations: Production Credit Associations which make operating and equipment loans, Federal Land Bank Associations which make real estate loans, and Agricultural Credit Associations which make operating, equipment, and real estate loans.

FCS organizations lend money from funds raised by the sale of bonds and notes. Most bonds are issued in denominations of $5,000 and typical maturity is six months. If you get an FCS loan, you are required to buy stock from between two and 10 percent of the amount borrowed. You get the stock back as the loan is repaid. There is a loan fee.

Once your long-term performance is established, only annual financial statements and crop progress reports are required for another loan and the credit check and verification are not as involved as for a new borrower. This reduction in paperwork reduces the costs of making the loan. You don't have to pay a loan fee each year. A revolving line of credit for a production loan can be made for up to four years. A multi-year production loan can be a problem, however, because you cannot be flexible in what you plant.

Analyzing Your Financial Situation

Bankers measure the risk-carrying capacity of a business in several ways:

1. Current ratio =
   current assets/current liabilities

   The current ratio should be at least 1:1. This means that you have a dollar in liquid assets for each dollar you will need to pay in the next 12 months.

2. Working capital =
   current assets - current liabilities

   This is the ratio of outstanding debt to the debt-free portion of your assets.

3. Leverage =
   total liabilities/net worth = debt/equity

   When this ratio is greater than 1:1, the lender has more invested than the borrower.

Association offices are located throughout California. A list is at the end of the chapter.

Farmers Home Administration

FmHA, an agency of the US Department of Agriculture, makes and guarantees loans to farmers who are otherwise unable to obtain credit. To qualify, you must be a permanent resident alien or US citizen, operate a family-size farm, have a satisfactory credit history, and have the education, training, or experience to operate a farm. The loans are to buy, expand, or improve operations. Some funds are targeted for socially disadvantaged operators who are African American, Native American or Alaskan Natives, Latino, Asian American or Pacific Islanders.

Eligibility is determined by the local FmHA county or area committee. Once eligibility is established, the county supervisor helps you design a farming plan. FmHA provides technical advice in carrying out the plan. FmHA also provides financial counseling.

If you qualify for a direct loan in all ways except cash flow, you may qualify for the "limited resource interest rate," usually three percent less than the going rate. The loan guarantee program includes a similar provision called "interest assistance." FmHA can subsidize interest payments up to four percent if it will improve cash flow to the point of making the loan acceptable to a bank.

FmHA guarantees loans for up to 90 percent of the loan and interest. Repayment terms range from one to seven years for direct loans. You negotiate the term and interest rate for guaranteed loans with the lending institution. The limits are $200,000 for direct loans, $300,000 for guaranteed ownership loans, and $400,000 for guaranteed operating loans. The maximum repayment term is 40 years.

FmHA will require you to pledge all assets as collateral. You must have a positive cash flow for a direct loan and a 20 percent margin for a loan guarantee. (A commercial lender generally requires a 30 to 40 percent margin on cash flow.)

There are state, district, and county FmHA offices in California. Obtain an application from a county office, usually located in the county seat. (Some offices serve more than one county.) Offices are listed in the phone book under United States Government. Also see the list at the end of the chapter.

■ Regional Development Corporations

Regional development corporations make loans and guarantee loans to small businesses under the auspices of the California Department of Commerce, Office of Small Business. There are currently two in the state, California Rural Coastal Development Corporation (Cal Coastal) and Valley Small Business Development Corporation.

Cal Coastal serves the region from Santa Clara to Santa Barbara. Cal Coastal makes direct loans to farmers using the FmHA loan guarantee program. The primary purpose of their loan program is to assist farmers already in operation. Loans are available for production, real estate, and refinancing. The loans are guaranteed by FmHA and their eligibility requirements apply. Prior farming history is crucial for loan approval. Cal Coastal also guarantees loans from other lending institutions for up to 80 percent. Special emphasis is given to farm operations that will create or retain jobs. The maximum term is seven years and the maximum amount is $350,000. For more information, contact:

Cal Coastal
5 East Gabilan Street, Suite 218
Salinas, CA 93902

Serving the central valley (Merced-Bakersfield):
Valley Small Business Development Corp.
955 N Street
Fresno, CA 93721

While not one of the regional development corporations, another group, SAFE-BIDCO (State Assistance Fund for Enterprise, Business and Industrial Development Corporation) participates in the Small Business Administration and the FmHA loan guarantee programs. They serve the North Coast region:

SAFE-BIDCO
145 Wikiup Drive
Santa Rosa, CA 94503-1337
(800) 273-8637

■ Other Sources of Credit

Credit to farmers is available through merchandising transactions to finance operating expenses. Credit is available from sellers of inputs and buyers of agricultural production. Firms selling equipment, fertilizer, seed, pesticides or buildings usually provide credit through open accounts. The farmer purchases inputs without making immediate payments. The added credit service is a way of attracting customers.

For annual inputs there may be an interest range. Billing may occur in 30, 60, or 90 days. These loans are very convenient for farmers and require much less paperwork than loans from other types of lenders. A disadvantage is that the terms of the loan are less flexible than other sources of credits.

By far the most common type of trade credit is from equipment dealers. These loans require installment payments. Loan documentation is usually more complete and a more complete analysis of repayment ability is made.

Firms buying commodities on contract may offer advances to their growers. This is most common for commodities that are in high demand.

Credit from individuals is usually in the form of a land contract real estate sale. The buyer of the property has full use of the property while the seller retains the title. Often the terms of the loan are 10 to 15 years with a large balloon payment at the end. This arrangement provides the seller with a steady flow of income at a relatively high interest rate without the tax burden of a complete sale. The risk is relatively low because the seller retains the title until the payments are completed. The buyer is able to purchase land without going through an extensive loan process. The interest rate is generally below that available from other credit sources.

Some operating loans are made to farmers by other individuals. These loans are generally within families.

Farmers Home Administration California Offices

STATE:
194 West Main Street, Suite F
Woodland, CA 95695-2915

DISTRICT:
1900 Churn Creek Rd., Suite 119
Redding, CA 96002

777 Sonoma Ave., Room 213
Santa Rosa, CA 95404-4731

3137 South Mooney Blvd.
Visalia, CA 93277-7389

21160 Box Springs Rd., Suite 105
Moreno Valley, CA 92557

COUNTY:
Alameda, Monterey, San Benito,
San Francisco, Santa Clara, San
Mateo & Santa Cruz Counties
635 South Sanborn Road, Suite 18
Salinas, CA 93901-4533

Alpine, Amador, Calaveras,
Contra Costa & San Joaquin
Counties
1222 Monaco Court, Suite 28
Stockton, CA 95207-6790

Butte & Plumas Counties
463-L Oro Dam Blvd.
Oroville, CA 95965-5791

Del Norte & Humboldt Counties
5630 South Broadway
Eureka, CA 95502-2027

El Dorado/Mono/Nevada/Placer
& Sierra Counties
251 Auburn Ravine Rd., Suite 103
Auburn, CA 95603-4294

Fresno County
4625 West Jennifer Street,
Suite 126
Fresno, CA 93722

Glenn & Colusa Counties
132 North Enright, Suite C
Willows, CA 95988-2697

Imperial & San Diego Counties
1681 West Main Street,
Room 412
El Centro, CA 92243-2285

Kern County
5500 Ming Ave., Suite 155
Bakersfield, CA 93309-8490

Kings County
680 Campus Drive, Suite D
Hanford, CA 93230-9505

Mendocino & Lake Counties
405 Orchard Ave.
Ukiah, CA 95482-5090

Merced, Madera, & Mariposa Co.
2135 West Wardrobe Ave., Suite A
Merced, CA 95340-6490

Modoc and Lassen Counties
204 West 12th Street, Suite E
Alturas, CA 96101-3211

Riverside, Orange, & So. Los
Angeles Counties
45-691 Monroe Street, Suite 1
Indio, CA 92201

Sacramento & Solano Counties
65 Quinta Court, Suite D
Sacramento, CA 95823-4386

San Bernardino, Inyo, & Los
Angeles Counties
15028 7th Street, Suite 2
Victorville, CA 92393-3859

San Luis Obispo/Santa Barbara/
Ventura Counties
582 Camino Mercado, Suite 582
Arroyo Grande, CA 93420-1816

Shasta & Trinity Counties
3179 Bechelli Lane, Suite 109
Redding, CA 96002-2098

Siskiyou County
215 Executive Court, Suite C
Yreka, CA 96097-2692

Sonoma, Marin & Napa Counties
777 Sonoma Ave., Room 212
Santa Rosa, CA 95404-4799

Stanislaus & Tuolumne Counties
1620 N. Carpenter Road, Suite 47
Modesto, CA 95351-1153

Sutter, Yolo & Yuba Counties
1531-B Butte House Road
Yuba City, CA 95991-2293

Tehama County
2 Sutter Street, Suite B
Red Bluff, CA 96080-4388

Tulare (So. of Avenue 240)
County
1963 East Tulare Ave.
Tulare, CA 93274-3297

Tulare (No. of Avenue 240)
County
3135 South Mooney Blvd., Suite A
Visalia, CA 93277-7360

Farm Credit System California Associations

Central Coast Federal Land Bank Association & Central Coast Production Credit Association
111 South Mason, Drawer AA
Arroyo Grande, CA 93420

Federal Land Bank Association of Bakersfield and Bakersfield Production Credit Association
5555 Business Park South
Bakersfield, CA 93389

Northern California Federal Land Bank Association and Northern California Production Credit Association
130 Independence Circle
Philadelphia Square
School House Building
Chico, CA 95927

Federal Land Bank Association of Colusa and Colusa-Glenn Production Credit Association
310 Sixth Street
Colusa, CA 95932

Federal Land Bank Association of El Centro and Imperial-Yuma Production Credit Association
1415 State Street
El Centro, CA 92243-2834

California Livestock Production Credit Association
8788 Elk Grove Boulevard, Suite L
Elk Grove, CA 95624

Fresno-Madera Federal Land Bank Association
1240 West Olive Avenue
Fresno, CA 93728

Fresno-Madera Production Credit Association
1250 West Olive Avenue
Fresno, CA 93728

Federal Land Bank Association of Kingsburg
1580 Ellis Street
Kingsburg, CA 93631

Intermountain Federal Land Bank Association and Sierra/Nevada Production Credit Association
Northeastern California
255 W. Peckham Lane
Reno, NV 89509

Farm Credit Services of Southern California, ACA
4130 Hallmark Parkway
San Bernardino, CA 92407

Sierra-Bay Federal Land Bank Association and Sierra-Bay Production Credit Association
3984 Cherokee Road
Stockton, CA 95215

Federal Land Bank Association of Yosemite
800 W. Monte Vista Avenue
Turlock, CA 95381

Central Valley Production Credit Association
800 W. Monte Vista Avenue
Turlock, CA 95381

Federal Land Bank Association of Visalia and Visalia Production Credit Association
3010-A West Main Street
Visalia, CA 93291

Pacific Coast Farm Credit Services
8741 Brooks Road
Windsor, CA 95492

Sacramento Valley Federal Land Bank Association and Sacramento Valley Production Credit Association
283 Main Street
Woodland, CA 95695

Taxes

A number of county, state, and federal agencies administer and collect taxes from farm operators:

The County Tax Assessor determines the fair market value and the County Auditor computes the taxes on land, buildings, and equipment. They are listed in the County Government Offices section of your phone book.

The Franchise Tax Board, 1912 I Street, Sacramento, CA 95814; (800) 852-5711, has information on California income and corporation taxes and distributes state tax forms.

The Board of Equalization administers sales, timber, excise, fuel use, and motor vehicle license taxes, and provides guidelines for assessment of property taxes. If you sell agricultural products, apply for a sales tax permit from them at 1912 I Street, Sacramento, CA 95814; (800) 852-5711.

The Timber Tax Division collects taxes on timber when harvested, based on board feet cut and type of trees.

The State Controller's Gasoline Tax Refund Division refunds taxes paid on gasoline used in off-road farm vehicles. Get a form from them at 3301 C Street, Room 301, Sacramento, CA 95816.

The Employment Tax District Office has information on unemployment insurance tax, disability insurance, income tax, and on programs that provide tax credits to employers. They are at 7001A East Park Way, Sacramento, CA 95816.

The Internal Revenue Service collects federal income tax, social security tax, and unemployment tax. The IRS's "Farmer's Tax Guide," Pub. No. 225, will help you prepare your federal tax return. Other tax guides which may be helpful for farmers are "Self-Employment Tax," Pub. No. 533, and "Tax Guide for Small Business," Pub. No. 334. Call (800) 829-3676 to request the guides and forms.

IV
MARKETING

Display at Sacramento Summer Harvest Tasting

Freedom, creativity, and innovation have characterized the small farmer for decades. Those qualities have built the remarkably efficient agricultural industry of the United States. Those same qualities are essential for you to be a successful and dynamic marketer.

Agriculture has long been **production oriented**. Our objective has been to produce, then ask, "How can I get people to eat more wheat?" or "more lettuce?" But, focusing on the product is fatal in today's marketplace.

You need to be **marketing oriented**. You need to identify customer needs and desires. This approach can have profound effects on business operations. You might find yourself exploring new products or new forms of packaging or spending more time with consumers who influence what you produce.

The kind of marketing that makes a small-scale farming operation profitable today is **niche marketing**: finding out what customers need or want and providing it; not

Chapter author: Daniel W. Block, Professor, California State Polytechnic University, San Luis Obispo just presenting what you have and hoping they will buy. You are in business to serve your customers' needs, and those needs dictate the type and form of your products.

As you shift from being production oriented to being marketing oriented, you will become a strong and profitable producer who meets your business objectives.

Planning

Frequently, people have excuses for why they can't plan: "Things change too rapidly to make a plan useful," or, "I'm too busy!" If you have unlimited money, unlimited time, and unlimited customers, you don't need to do much planning. But all of us in small farming know our capital and time are limited, and we have to share our customers with an increasing number of competitors. In today's marketplace, if you are responsible for an agricultural operation and you do not have a marketing plan, you are playing with a time bomb that will surely go off within the next decade! Knowing your customers' needs and having a written program to satisfy those needs is vital. (Filling out the worksheets at the end of the book will help you develop your market plan.)

Developing a marketing plan can be a simple project that can yield tremendous, long-term benefits. A marketing plan allows you to maximize limited resources and to react to changes more effectively.

As the marketplace becomes affected by rapid change, increasing international competition, and decreasing resources, an interesting phenomenon is occurring. Whereas producers used to make a product and sell to any and all who could buy it, today's market-oriented farmers define their target market and focus their time and resources on that target exclusively.

Planning to focus your marketing energies on a specific target often means the difference between wasting valuable resources while hoping someone will buy your product and efficiently managing your resources to "get the most bang for your buck." Marketers call this "targeting a market segment" or "picking a niche."

You cannot be all things to all people. You can make more money, more efficiently, when you tailor your product and services to specific segments in the marketplace for whom you can do a good job. Finding your niche in the market means finding customers who have needs that you can satisfy better than anyone else. It means differentiating your product to a specific segment of customers, and building a relationship with those customers. They will perceive you to be especially qualified or equipped to satisfy their unique demands.

To get information you need to do marketing research. You need to know all about your customers and your competition. You need to know who your customers are, where they live, what they buy, how they buy, when they buy, and who influences their purchases. You need to know which customer needs are not being satisfied.

Marketing Research

Look around and see who is successful. A potato farmer in the Bakersfield area, seeing that other growers packed and identified their products as "premium" and got higher prices, designed his own box to differentiate his product. Buyers began to perceive his potatoes as having better quality, and he started to get better prices. Once he ran out of the new boxes and had to use a neighbor's. He received a reduced price even though the potatoes in the boxes were the same as he had always delivered to his customer. He was marketing perceptions as well as food.

The California Department of Food and Agriculture and UC Cooperative Extension distribute publications and sponsor seminars that will show you how to put together a marketing program for your operation. Attending trade association meetings and conventions is also helpful.

Today's small farmer needs to read metropolitan, national, and international newspapers to be aware of what's going on in the marketplace and with customers. Your local newspaper is not enough.

Divide up all the customers in your market into categories. Customers can be defined by geographic areas, such as San Francisco, Ventura, or cold climates and warm climates. Customers can be defined by their lifestyle—energetic or sedentary. They can be defined by age, such as retired people, young families, teenagers, or children. They can be defined by taste preferences—some like things sweet or tart, large or small, processed or raw. They can be defined by season or time, such as spring or fall calving, pre-planting or post-emergence. Precisely identify the unique needs of the customers in your market, and select those needs which you can most uniquely meet and for which there are large enough numbers of customers for you to be profitable. This gives you the edge over your competitors and conserves your resources.

Another form of market research is simply to ask your customers what they want. Conduct a personal or telephone interview or send out a simple questionnaire to people with whom you deal. If you own a roadside stand, ask people who shop at your stand why they stopped, what they are looking for, and what else you can do to satisfy them. If you are thinking of selling to the food service industry, such as restaurants, ask the chefs what they want and in what form they want it. You will become aware of opportunities you never before considered.

All of these ways of defining customers provide a way of finding potential opportunities in the marketplace that are not always apparent. Agricultural producers who have been successful in targeting specific niches include vegetable growers who provide Asian vegetables to the expanding Asian populations in southern California; or producers who target restaurants and provide them with products of specific sizes and quantities; or alfalfa hay producers who grow exclusively for the purebred horse market; or a cherry producer in Stockton who packs cherries specifically for the very particular tastes of his customers in Hong Kong. All of these people are specialists to their customers, uniquely suited to meet their customers' special needs.

Situation Analysis, Objectives, Strategies (SOS)

Budget Action-Plan Measurement (BAM)

Situation Analysis: Where are you now?
- Market potential.
- Customer needs.
- Differential advantages of your product.
- Competition's strengths and weaknesses.
- Your operation's strengths and weaknesses.
- Possible market segments or niches.
- Industry trends.
- Pick a target market.

Objectives: Where are you going?
- Must be measurable.
- Must have a completion time or date.
- Must be specific.
- Must be attainable.

Strategies: How will you get there?
- What product does your customer want and in what form will it be?
- How will you distribute/sell to your customer?
- What price should you charge?
- How will you promote it (advertising, personal selling, public relations, publicity or special incentives)?

Budget
- What will these strategies cost?
- What will be the financial return?

Action-Plan
- When should you do the recommended strategies?

Measurement
- Are you making progress toward your objectives?
- Did you achieve your objectives?

Many small farmers are doing these things already. They are getting an advantage over those who still spend all their time in the field. If you aren't already doing so, get off your tractor and go talk to your customers!

SOS/BAM: A Tool for Developing a Marketing Plan

A simple structure for developing a marketing plan is the SOS/BAM model. This structure, when used properly, will give you an operational tool that is as handy as a pair of pliers or a screwdriver.

The **situation analysis** is the most important part of your marketing planning and requires careful thought. You analyze where you are right now—your situation. You look at market potential, customer needs, your product's and your operation's strengths and weaknesses compared to your neighbor's, industry needs. You talk to your neighbors, observe operations that are successful, consider articles and journals you've read that tell about trends in your area of business. Analyze your market for potential niches. Ask your employees what they think your operation's strengths and weaknesses are. You might be surprised at what they say.

You don't need to accomplish this analysis in one sitting. It may take a few periods of writing separated by intervals of thinking and information gathering. An almost magical process begins to take place when you do a situation analysis. New ideas emerge. You'll discover creativity in yourself and your employees you may never have known were there.

The **Objective** section is very simple. You put down on paper specific, measurable objectives you'd like to achieve with your operation. You may want to increase the number of customers, or want people to be more aware of your brand name, or want a greater profit. Good objectives are measurable and have a completion date. Don't say, "I want to sell more watermelons." Say, "I want to sell 10 percent more watermelons by August 31." Make your objectives attainable. They should make you stretch, but not be impossible to achieve.

Strategies is the part of your plan where you put down ideas for increasing your business through marketing. Include:

- Product development—packaging, branding, warranty, and service to meet the unique needs of your target market.

- Pricing decisions—raising your price or asking for a premium price.

■ Place or distribution decisions—deciding whether to continue with your present buyers.

■ Promotion—advertising, personal selling, public relations and publicity, special incentives; communicating to your customers that you can satisfy their needs.

The BAM in the model stands for:

Budget—there has to be an economic justification for implementing your plan.

Action-Plan—calendar of events stating when you are going to start these activities.

Measurement—a means of evaluating your progress to see if you are on your way to achieving these objectives. A marketing plan is essential in the competitive market today. It is not an overwhelming task, and it yields some interesting results. Developing a marketing plan is a catalyst to new and innovative ideas. If you make the effort to begin the process, you'll find yourself thinking in directions you never before considered.

Kozlowski family displaying jams and jellies

Product Strategy

Quality

Your most important marketing strategy is to produce the most desirable product you can. All other marketing factors—advertising, public relations, farmers' markets, roadside stands, salespeople—are useful only if your product truly satisfies the customers and is of the quality that you claim it is. The old saying that began in agriculture is appropriate today: "You can't make a silk purse out of a sow's ear!"

But, remember, working 12 hours a day in the field to produce the highest quality product you can is not all there is to it. Beyond the physical features of your products, which you work so hard at perfecting, lie features that are less visible yet eagerly desired by your customers, and often neglected by producers.

Adding Value

You must differentiate what you are offering from all other products in the marketplace to show that your product will better satisfy the needs of your target market. This is called "adding value," and the more value you add to your product, the more profit you will receive.

One way of adding value is with a **guarantee**. This can set your product apart. An example is a Red Angus purebred breeder in Montana. He guarantees his heifers will be free from calving problems. If a rancher has calving problems, his money is refunded or he gets a new heifer. This breeder's customers are purchasing the security of knowing they have minimized their financial risks. He is selling peace of mind.

Another way of adding value is with a **special phone number** or hotline. Some marketers have a toll free number for their customers.

Adding **information and education** to your product is always of high value to your customers. Consider seminars or workshops. Information about cooking it, processing it, delivering it, cutting it, vaccinating it, unloading it, or boxing it are a few options that may apply to your business. All will add value to your product.

Building a **close relationship** is one of the oldest and most effective means of adding value to your product. If two carloads of grain are the same quality, the same price, and can be delivered at the same time and under the same terms, the seller who has the closest and most helpful working relationship with the buyer will get the business.

If you can **keep the product longer**, you add value. Perhaps you can dry, ferment, or can it. You may produce jams, jellies, preserves; juices; cow, goat, sheep milk and cheese; pies; vinegars; smoked meats; wreaths; garlic braids, etc.

Perceptions

A product is more than its physical and functional characteristics. Your customers are not simply purchasing material items with characteristics resulting from your combining soil and water, or genetics and feed. They are also purchasing a perception of more value for their money, whether it be through services, guarantees, or even an image.

A farmer with 80 acres of apples in Tehachapi markets his blossoms. At blossom time, he hauls Los Angeles area senior citizens through his orchard in reconditioned manure spreaders. They love it—it's a refreshing change from the congestion of the city. He's selling an experience, an image, a perception. It's a simple marketing technique and it's fun. He gets about 300 busloads a year, and he makes a profit. That's understanding what your product really is.

Brand Name

Another way to differentiate your business or product is by using a brand name. Many producers, both large and small, identify their products with a brand name or logo. This identification can be used on a box, on advertising, on packaging, or on the product itself. Studies show that both domestic and foreign customers prefer branded produce over unbranded. They think it is of higher quality.

When selecting a brand name, pick one that will mean something to the customers, that will make them want to buy your product. Keep it simple to say. It's usually a waste of a good marketing opportunity to pick a name that you think is clever or has your family name as part of it. "Butterball" says nothing about the farmer who grows the turkeys, but certainly gets the customers licking their lips.

Pricing

The old adage that "farmers are price takers, not price makers" is not always true today.

If you want to receive a higher price for your product, you have two alternatives: 1) get the government to guarantee a higher price through price supports or subsidies, or 2) add value to your product. You probably won't choose the first alternative because it is very difficult.

Providing customers with a quality product—a product with the features that the customers want, both tangible and intangible, allows you to ask a higher price.

A large citrus grower in Porterville garners a higher price than his competitors by providing high quality produce **on time**. The large chain store to which he sells pays a premium price because they know the order will be consistently filled with high quality produce, on time, without problems.

Place

Out of financial necessity, the battle cry of the innovative producer today is, "Get close to your final consumer!"

Today it is much easier than in times past. For decades it was almost impossible for a farmer to communicate with the consumer because it was a long way to town and there was really no reason for communication. Farmers were making money, folks in the city were getting their food, people had respect for farmers, and most producers were too busy working on the farm to worry about a system that was working satisfactorily. Producers relied on the railroads and vast networks of brokers, wholesalers, grocery store owners, and small vendors. Most people were sympathetic to problems of farmers.

But that situation has changed. Farmers are struggling to find a profit. Although food continues to flow smoothly to the 98 percent of the US population that is not on a farm, a lack of sympathy for how food is produced has created a relationship problem between the farmer and the consumer. One thing hasn't changed: the farmer is still working hard. But, working hard is not a valid excuse for not getting involved in the marketing process. Communication and transportation today allow you to bypass middlemen. You can sell directly through roadside stands, farmers' markets, or to local markets and restaurants. You can even make contacts in Tokyo, New York, Atlanta, or Hong Kong by spending a few dollars on a phone call. For a relatively small investment in a fax machine, you can communicate globally for just pennies.

Bypassing traditional distribution channels is appealing. Nevertheless, remember, the sword that decapitates the middleman is two-edged. An ill-planned strategy could overlook some vital functions your middleman performs, and which you may not be able to duplicate. Developing the necessary contacts and maintaining the retailer and jobber relationships in markets far from your operation take time and money.

Promotion

Promotion is what marketing is all about. You want to let customers know that you have the product with the features and benefits they need. Reaching the largest number of people in your target market for your dollar is your goal, and the method of promotion you select should be governed by this principle.

Advertising is the most visible form of marketing, but promotion also includes personal contact, public relations and publicity, direct mail, and special promotional incentives.

Personal Contact

The oldest and least expensive form of promotion is "word of mouth." That form of customer goodwill and support cannot be bought with flashy advertising. It is

Intensive Vegetable Crop Production

By Stephenie Caughlin,
San Diego County farmer

I have 1.5 acres. I pull 1,000 pounds of produce a week, 365 days a year. My place is a miniature farm. I have 200 chickens, a small duck flock, and just over 40,000 square feet under cultivation. The duck eggs are sold directly to the public for the people who are allergic to chicken eggs. The duck eggs are pre-sold for $4.25 per dozen; the price is reasonable considering the labor and other inputs. People buy them before I even get them to the market.

My farming system is above-ground raised beds with an organic approach. When I first started farming, I figured out the square footage and number of times per year that I could turn a crop. If you get a dollar per square foot with 40,000 square feet, you're looking at $40,000. At fifty cents per square foot, income decreases to $20,000 a year and at seventy-five cents per square foot, you are at $30,000. If you push it to the upper limit, $1.25, then your income will reach $50,000 a year.

Quality is most important for income. Consumers may be willing to pay more for high-quality products. The produce doesn't have to be picture perfect but it's got to be at the peak of its flavor and it's got to look good when you present it. Most of my business is done with farmers' markets. At supermarkets I wholesale lettuce at $.75 to $1.00 a head. They mark it up to between $1.49 to $1.89. They buy everything I present to them. Some grocery stores are not allowed to buy direct, but you won't know unless you ask. I haven't had experience with restaurants because I haven't really made an effort to sell to them. I already have my markets in place, but I understand farmers have done extremely well selling directly to restaurants.

I sell "mesclun"—a mix of 26 varieties of lettuce. I call it a country mix. People like getting something new and different, but don't like pretentiousness. Sometimes specialty food approaches go too far. Make a joke, put the customers at ease. If you are introducing something for the first time, call it yuppie chow. They laugh and relax. They are not as intimidated.

Another cute idea is the way we sell tiny bunches of carrots. I use heirloom carrot varieties, and crowd them in a bed so a lot of carrots twist around each other. I can't sell them to a grocery store. But if I put them out on my farmers' market table and call them "love carrots," people think it's the cutest thing in the world and I can charge double!

earned by delivering the highest value to your customers for their money. For people making buying decisions, nothing beats a person-to-person relationship between buyer and seller. A vegetable grower in the Bakersfield area makes regular trips to his buyers in the Los Angeles area just to say hello. He knows that a smiling face associated with his farming operation goes a long way to differentiate his produce from his competition's.

There's always room to strengthen your personal relationship with your customers. One rancher sends birthday cards to an extensive list of buyers. Another producer telephones his buyers regularly. The power of personal relationships lies not only in words, but also in non-verbal communication. Customers sense when you sincerely want to satisfy their needs. The world has become so computerized that your personal "How can I help you?" will be a breath of fresh air.

Advertising
Dancing raisins, avocados wearing party hats, musical videos showing the latest "models" of Angus bulls, and colorful direct mail catalogues attest to the fact that agricultural producers in the US are awake to the potential of sophisticated advertising. Advertising agencies throughout the country are busy putting together clever campaigns to promote turkeys, almonds, milk, specialty fruits, cattle, and dozens of other agricultural commodities. This trend will continue as producers strive to differentiate their products and garner higher margins.

Chefs and producers at Sacramento Summer Harvest Tasting

When deciding how and if to advertise, find out which media your target market reads, watches, or listens to. Don't throw away money placing your ad in the wrong type of media. Most reliable publications and stations will give you a free profile of their subscribers or listeners. Then you can decide if that publication or station is the best way to reach your target market.

Direct Mail

Direct mail offers an effective way of communicating your message. Many purebred cattle producers send out catalogues and mailers to ranchers to promote their sales and maintain customer loyalty.

Successful Marketers

Successful marketers are more concerned with what their customers want than with what they are producing. They are listeners and question askers. They are long-range thinkers and are willing to forego short-term profits for long-term growth. They are curious about the world, and hungry to learn. Successful agrimarketers are identical in one area: they like people and they enjoy serving them. That is what marketing is all about.

Additional Resources and References

Fresh Trends: A Profile of Fresh Produce Consumers (annual edition). Overland Park, KS: Vance Publishing Company.

Levinson, Jay Conrad. 1984. Guerrilla Marketing: Secrets for Making Big Profits from Your Business. Boston, MA: Houghton Mifflin Co.

Marketing US Agriculture (1988 Yearbook of Agriculture). Washington, DC: US Government Printing Office. 327 p.

Nelson, Theodore. Measuring Markets: A Guide to the Use of Federal and State Statistical Data. 1987. Washington, DC: US Department of Commerce, Industry and Trade Administration.

Smallwood, M., J. R. Blaylock, and J. M. Harris. 1987. Food Spending Trends in American Households, 1982-84. Statistical Bulletin 753. Washington, DC: US Department of Agriculture, Educational Research Service.

Tourism USA: Guidelines for Tourism Development. 1986. Columbia, MO: University of Missouri.

V
SELLING YOUR PRODUCT

Introduction

You can market your products in many ways, including through:

Exporters
Brokers
Wholesalers
Certified farmers' markets
Cooperatives
Roadside stands
Fairs and flea markets
Shippers

U-pick operations
Rent-a-tree arrangements
Gift packs, mail order businesses
Buying clubs
Community supported agriculture plans
Retail stores
Restaurants

Before you begin production, decide which methods you will use. Consider your contacts, types and accessibility of markets, market conditions, the amount of time and money you have, and the risks you are willing to take. Talk to marketing agents to determine who is interested in buying, handling, or marketing your products. These agents are shippers, wholesalers, brokers, farmers' market managers, restaurant chefs, etc. Proximity to a large metropolitan area provides a great advantage for marketing.

Chapter authors: Stephen Brown, former Marketing Advisor, Los Angeles County Cooperative Extension; Claudia Myers, Associate Director, Small Farm Center, UC Davis; Louie Valenzuela, former Limited Resources Farm Advisor, Santa Barbara County Cooperative Extension; David Visher, Program Representative, Small Farm Center, UC Davis; Paul Vossen, Farm Advisor, Sonoma County Cooperative Extension

Marketing is divided into two broad categories: conventional or traditional marketing and direct marketing.

In **conventional marketing**, you pay "middlemen" for handling facilities and services:

- Box and crate assemblers provide standardized containers. Many fruits and vegetables have container regulations. Also, if you want to market under your own label, box and crate businesses can provide art and design services.

- Loading docks are used for unloading and reloading onto refrigerated tractor-trailers.

- Cooling facilities are usually provided by wholesalers until the shipping destination is determined. At cooling facilities, products are put on pallets and reloaded into refrigerated trucks.

- Trucks—common carrier transporters (Public Utilities Commission regulated)—transport and deliver your product to a wholesaler in a terminal market, usually in a large metropolitan center.

- Terminal markets are where buyers and sellers meet to make their deals.

Direct marketing removes the middlemen and puts you, the farmer, in charge of the whole process—from planting to consumer purchase. Direct marketing requires time, effort, money, personal connections, and information on marketing conditions. In return you receive a higher price per item sold. You must arrange for:

- Picking/gathering
- Sorting
- Packing
- Cooling
- Shipping
- Retailing
- Brokering
- Storing

Consider all types of marketing practices. An increasing number of small farmers are investing in more than one marketing approach. By doing this, they buffer their risks in the competitive marketplace.

Market Windows

When a product is marketed is critical. Determining when to market involves gathering information from sales people, produce "bird dogs" (individuals who act as the eyes and ears for sellers or buyers), field agents, wholesalers, marketing board staff members, farmers' market managers, etc. Most vegetable farmers attempt to produce for a "market window." Market windows are opportunities in production and marketing not yet fully exploited by other growers. Marketing windows may be "supply gaps" as recorded by the office of the Federal-State Market News Service. The Market News Service publishes daily reports for commodities showing historical price and supply descriptions of crops currently being marketed. Correlating the information received from the Market News Service with the

Mariani Orchards: Diversified Marketing Strategies

by Erin Chapman, Agricultural Consultant

The Mariani family has farmed in Morgan Hill since 1958. Over the past 30 years, Andy Mariani has seen many family farms go out of business. In 1958, there were 25 successful orchards in the area; now there are only two. A combination of factors induced this change: farmers grew too old to farm and relatives weren't interested in carrying on the farms, or prices were too low and production costs too high, or urban development and property taxes increased. One reason Mariani Orchards has survived is because they diversify production practices and marketing strategies.

The Marianis own 60 acres of land and lease another 40 acres. They grow numerous varieties of stone fruits. One of their specialties is the rare Muir peach. It is heart shaped and golden yellow. This delicate peach must be handled carefully after harvest. Andy said, "At the turn of the century, there were more Muir peaches in California than any other variety. It was a standard drying peach because it's sweet and dries well. Now I don't think there are more than five commercial orchards producing Muirs in California."

About 20 percent of all the fruits sold by Mariani are actually grown on the farm (cherries, apricots, peaches).

The rest (pears, figs, prunes, nuts, dates) are bought elsewhere. The cherries are primarily sold to wholesalers in Los Angeles, San Francisco, and Oakland. Some cherries are shipped to Taiwan and Japan.

When the fruit is ready for drying, it is halved, pitted, and placed on wooden racks to dry. Sulfur is added by enclosing the racks with plastic sheeting and burning sulfur under the tent. The pears are sulfured for three days, the apricots, six hours. "We don't make money by simply producing apricots, we make money by drying, processing, and selling them." After drying, most of the fruit is processed and packaged in San Jose. About two percent of the dried fruit is saved for selling directly to customers at the farm store, and five to 10 percent is reserved for mail order sales and special orders. The rest of the dried fruit is sold to distributors, wholesalers, and retailers. Future marketing plans include creating special gift packages for direct sales.

By selling directly to the consumer, Andy receives the retail price, which is about 200 percent higher than would be received for dried apricots, and 150 percent higher for dried peaches sold to a packer. "We expand every year; our ultimate plan is to have a large retail outlet on the farm."

experiences of various buyers helps growers determine the profitability of early and late production, exotic crop production, and traditional seasonal production.

Market windows are of two broad types: out-of-season production of commonly grown fruits and vegetables; and production of specialty, exotic, or uncommon products.

Out-of Season Products

In California, climatic diversity has created new opportunities—such as early vegetable production in the desert region or late fruit production in the coastal valleys. Of course, as soon as a large number of growers identify a market window, the opportunity is lost.

Some growers extend their seasons of production by creating an environment that does not naturally occur in their areas—usually by using greenhouses or plastic tunnels.

Specialty Products

Of increasing importance is the production of exotic or unusual varieties of fruits and vegetables. Just as other market windows have been lost, these marketing opportunities will be also as competition from other growers increases. Selecting which crops to grow also depends on affordability and quality of labor and equipment.

Opening the windows of opportunity is never ending and is the key to successful marketing for growers of all size.

Marketing Orders and Commissions

Marketing orders and commissions are set up to aid in marketing some commodities and establishing standards for size, grade, and/or maturity. There are federal and state marketing orders and commissions. Some assess fees to growers to pay for research, advertising, or promotion. All the crops or products listed below are affected by a marketing order or commission. Some of them grant an exemption for direct marketing or for selling small quantities. For more information, contact the California Department of Food and Agriculture Marketing Branch in Sacramento at (916) 654-1245 and ask for the manager of the appropriate order or commission.

Alfalfa	Citrus	Melons	Plums
Almonds	Dates	Milk products	Potatoes
Apricots	Dried figs	Nectarines	Poultry
Artichokes	Dry beans	Olives	Prunes
Asparagus	Eggs	Oranges, Navel	Raisins
Avocados	Flowers	and Valencia	Rice
Bartlet pears	Grapes	Peaches	Strawberries
Cantaloupes	Iceberg lettuce	Peanuts	Tomatoes
Carrots	Kiwis	Peppers	Walnuts
Celery	Lemons	Pistachios	Wheat

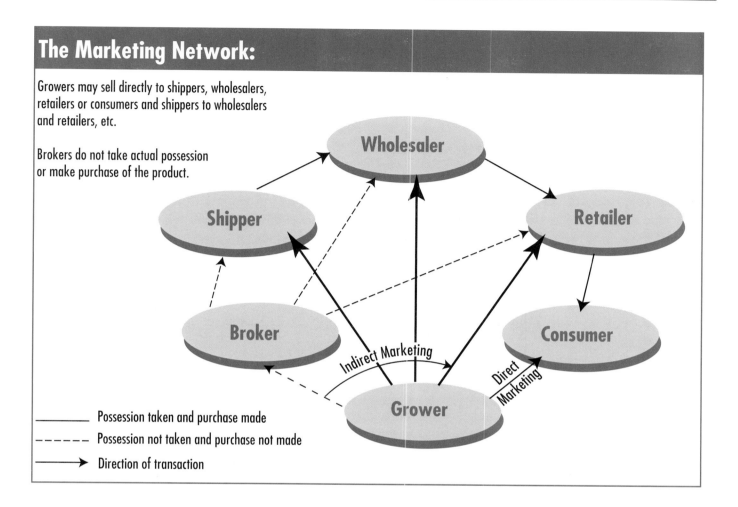

The Marketing Network:

Growers may sell directly to shippers, wholesalers, retailers or consumers and shippers to wholesalers and retailers, etc.

Brokers do not take actual possession or make purchase of the product.

———————— Possession taken and purchase made

– – – – – – Possession not taken and purchase not made

——▶ Direction of transaction

Conventional or Traditional Marketing

Wholesalers

Many wholesalers possess years of marketing experience and numerous contacts. Unless you are willing to put forth extraordinary efforts in making direct and continuous contacts with retailers and consumers, you'll want to consider using wholesalers.

Produce wholesale marketing is competitive and rapid due to the large number of handlers and large number of items for sale. There is competition among the many sellers, shippers, handlers, and buyers. There may also be competition between products that can serve as substitutes for one another. For example, the price of peaches will affect the demand for nectarines.

Trade newspapers, such as The Packer and Produce News, provide information on national and international marketing trends that may be predictors of future prices. The Blue Book and The Red Book are two regularly updated industry credit rating "bibles" for buyers and sellers. They contain names and addresses of shippers, buyers, truck brokers, etc. from across the nation. Instructions on handling claims, rules of transport, state laws and regulations are also covered. There are subscription costs for The Blue Book and The Red Book.

Produce News Weeklies

The Packer
P.O. Box 2939
Shawnee Mission, KS 66201
(800) 255-5113

Produce News
2185 Lemoine Avenue
Fort Lee, NJ 07024

Produce Directories

The Blue Book
The Fruit and Vegetable Credit and Marketing Service
Produce Reporter Company
Wheaton, IL 60187

The Red Book
7950 College Blvd.
Overland Park, KS 66210-1821

Terminal markets are colorful, bustling centers of free market enterprise where buyers and sellers meet. Information on quality and quantity of items available for sale is passed around quickly. For growers, produce wholesalers at the terminal market provide access to retailers (grocery stores, supermarkets, restaurants, hotels, other produce wholesalers). The wholesaler, acting for the retailer, helps the grower out by giving information on consumer eating habits. The grower can often schedule time and type of production on the advise of the wholesaler.

The terminal markets of Los Angeles and San Francisco give growers access to local, national, and international clients. Some wholesalers cater to small supermarkets, where small growers can market low volume products. Large supermarket chains often buy directly from large growers.

Restaurant purveyors buy for chain and individual restaurants. Small growers will find it advantageous to contact restaurant purveyors because they can plan production for the specific needs of these well paying clients and the volume required is often not demanding.

Selecting a Wholesaler. Depending on your quantity and types of crops, you may work with one or more wholesalers. Some houses specialize in commodities, while others carry a full line of products. Selecting more than one buyer will increase your selling chances in a competitive market. However, having a good working relationship with one dependable buyer is far better than securing the service of many unreliable buyers. There are several methods for selecting a wholesaler:

- Word of mouth—the reputation of a wholesaler who has been in business for many years will be well known.

- The Red and Blue Books—use the listings of the watchdogs of the produce industry.

- Visit a terminal market and talk to buyers.

- Select a buyer by phone when rapid sale of your merchandise is desirable. Once you select a buyer, conversing by telephone will be the standard method of communication.

Wholesaler's Expectations. Once you have selected a wholesaler and your product is ready for delivery the wholesaler will expect:

- Seven to 10 days notice on large deliveries or one day notice on small deliveries.

- Items packed in standardized containers.

- Items of high quality.

- Amount to be shipped as agreed upon before shipment.

- Price per weight of your product should be negotiated and determined the day before shipment. Wholesalers sell your products on a consignment basis—you assume all price and product risks until the product is sold.

- Supply of items should be consistent over the season.

Retail Promotion via the Wholesaler. Buyers often request seven to 10 days notice before delivery of large loads so they can arrange with retailers for shelf space or for radio and newspaper advertising. Cost of advertising is usually passed back to the grower. Promotional advertising is needed if:

- There is an oversupply of products.

- Seasonal products have arrived.

- An item is discounted at the retail level.

- Quick sales of highly perishable items are needed.

Wholesaler Assisted Finance. Once the grower and wholesaler have developed a solid working partnerships, the wholesaler may help finance production and, in the event of crop failure, share in the financial loss.

Standardized Packing. Whether you pack yourself or send products to a packing house, use standardized containers. Standardized containers contain an established number of the items at a predetermined minimum weight per container; this helps clarify the price per pound. Not using standardized containers may lead to significant miscalculation of the money to be exchanged. Each commodity has a specific or several specific box sizes and weights that can be used.

Packaging standards are established by USDA and CDFA and enforced by the County Agricultural Commissioner Office or Market Enforcement Office (California Department of Food and Agriculture). These offices are often located in the terminal markets.

Many specialty commodities are not yet assigned official containers but are sold in "agreed-to" containers—determined by the trade. Contact your wholesaler or Market Enforcement Office before packaging specialty products.

Price and Payment. The price of products is determined by grade (US Extra Fancy, US Fancy, etc.) and the quantity entering the market. An example of a typical price breakdown is as follows:

A standardized box of 80 Washington delicious apples (40 to 42 pounds net weight, 45 pounds gross weight) is sold by the grower at $18.00 FOB (freight on board or cost without transportation included). Freight cost is an additional $2.00 per box to be assumed solely by the buyer or by both grower and wholesaler. The wholesaler sells the box to the supermarket at $23.00 per box (usually no more that an 18 percent profit). The market sells the apples at $.99 per pound (as much as 100 percent mark-up). If a broker (a broker brings buyers and sellers together and facilitates title transfer) is involved in the transaction, subtract an additional $.15 to $.35 per box from the amount received by the grower.

If the supermarket refuses the box of Washington delicious, responsibility for the loss is the grower's. However, the wholesaler usually will attempt to sell the product to another buyer. In that case, price adjustment may be necessary and is usually due to lower product quality.

Payment to the grower is usually made by the wholesaler within three weeks after receipt of the product.

USDA Inspection. Price paid to growers by the wholesaler is dependent on quality and quantity of the product. A grower may ship US Extra Fancy Washington delicious apples, but, upon inspection, the buyer may not agree with the stated grade. In this case, either or both parties may request the service of the USDA Fresh Product Inspection Branch. The inspector will view the product and determine if the shipment meets the minimum requirement as labeled. Determination is based on the quality and size of the produce. The shipment either passes or fails the stated grade. There is no re-grading. A failed grade means re-negotiation of the price or rejection of the shipment. The service of the Fresh Produce Inspection Branch is available only upon request to sellers and buyers of fresh produce.

Shippers

Some growers sell their products directly to buyers through the services of shippers. Shippers locate buyers (normally chain stores or wholesalers in major metropolitan areas). As in selling to wholesalers, similar marketing protocol follows: shipper and buyer agree to a price; shipper arranges and pays for transportation, cooling, or special handling; shipper charges a commission; shipper pays the grower the net price after all deductions are made. A price adjustment may be needed to deal with unsatisfactory quality. Shippers may assist with production cost and may agree to suffer financial loss for unsatisfactory crop yield. Shippers may grow their own product and consolidate with other growers to augment their market share.

Direct Marketing

Farmers' Markets

At certified farmers' markets (CFMs) farmers sell their products directly to consumers. CFM locations are approved by County Agricultural Commissioners. Products do not have to be in standardized packs. A CFM may have as few as five growers with total annual sales in thousands of dollars, or, in a large city, as many as 100 growers with total annual sales in the millions.

In 1992, there were over 180 CFMs operating in California and approximately 3,000 producers selling at them. There was an average of 33 growers selling at each market.

Farmers' markets operate on specific days at specific times. At popular markets during summer months, stall space is limited. Many markets establish priority rules, stall reservations, systems for tracking attendance, and waiting lists for future stall vacancies.

Most markets charge a weekly fee for growers to participate. At about a third of the markets, the fee averages $17 a week; other markets charge a percentage of gross weekly sales (an average of six percent).

A Certified Producer's Certificate gives you the right to sell fresh fruits, nuts, vegetables, eggs, honey, flowers, and nursery stock directly to consumers at CFMs without the usual size, standard pack, and container and labeling requirements. These exemptions save the grower time and money.

You can obtain a Certified Producer's Certificate from your County Agricultural Commissioner. You need to be actively working the land that you own, rent, lease, or sharecrop.

You may also sell "non-certifiable agricultural products." Check with the farmers' market manager to find out what is approved. Typically these items are: processed products like shelled nuts, jams and jellies that have been produced or derived from plants produced by you; catfish, trout, and oysters from controlled aquaculture operations; livestock and livestock products, poultry and poultry products derived from animals raised by you. All state and local health regulations must be followed.

Sources of CFM Rules and Regulations

California Code of Regulations, Title 3, Food and Agricultural Code, Article 6.5, Direct Marketing.

California Health and Safety Code, Chapter 4, California Uniform Retail Food Facilities Law (CURFFL).

California Sherman Food, Drug, and Cosmetic Law, Division 21, Section 26000.

Individual CFM rules.

Farmers who are successful selling through farmers' markets either have a diverse mix of products or they produce something that contributes to the overall mix of items for sale at the market; they have sufficient and continuous supplies throughout the season or year; and they participate in the market regularly.

Find out about the CFMs in a particular county from your County Agricultural Commissioner. You can also contact Southland Farmers' Market Association, 1010 South Flower Street, No. 402, Los Angeles, CA 90015, for a complete list of the markets in the state.

Selling Fresh Fruit at Farmers' Markets
By Art Lange, Honey Crisp Farms, Reedley, CA

There is a good demand for tree ripened fruit. The volume is small in the total picture but growing. Certified farmers' markets present a good opportunity for me, as a small grower, to sell my fruit.

I'd like to share some ideas that have helped us :

Our trees are planted close together because it saves on labor and land. We keep our trees short and open to light. We fertilize, irrigate, and thin heavily to get larger fruit. We use old high-quality varieties like Hale, Elberta, and Santa Rosa, as well as new, good tasting ones.

We constantly test old and new varieties. We just planted Elephant Heart plum. We grow about 150 varieties of trees and vines, but sell in quantity only about 20 of these. To stay in the market from mid-May until mid-September, we need peaches, plums, and nectarines that ripen at different times. One-half acre of a variety is the maximum I can handle. I try to have about 24 trees of each variety. If it is one of my best varieties, Snowqueen nectarines and Springcrest peaches, I will go up to 200 trees which is equivalent to one-half acre of standard plantings in production amounts.

We work 14 to 18 hours a day. If you are not willing to work long hours, you may want to think carefully about selling at farmers' markets.

We pack in the field. We do not wash our fruit. We do not machine size, grade, or defuzz, so no one handles the fruit after it is picked until the customer buys it. We pack all our fruit in single layer boxes and transport the most fragile to the market on four- to six-inch Styrofoam pads. This seems extreme, but it give us the quality we seek.

We do not store our fruit, but cool it in a small cold room at 45° F. This makes it last a few more days in the market.

We arrive at the market early to set up. We let the manager know the quality of the produce we have for sale. It is important to the market and to us that the market have good-quality produce. There are many farmers who want to sell at farmers' markets but want to sell only their culls. This does not help build markets, because the quality is not good enough.

We sell our products at a high price. Price is only one of the factors consumers consider in deciding whether or not to buy. Product quality, freshness, flavor, and bright attractive colors are more important with our customers. Our peaches, nectarines, and plums are top quality. We use the prices charged by other growers as our starting point. We take into account how the quality of their produce compares to ours. Where there is competition from many local growers and homeowners with their own trees and where fruit is plentiful (like in Fresno), we are lucky to sell $50 worth in a day, so we travel long distances. We sell our fruit away from the Central Valley in places like San Francisco, Santa Monica, and San Diego.

Give out samples of your fruit for on-the-spot tasting if the market allows. It will cost you money to supply this fruit, but it will educate consumers very quickly about what they are buying.

In five years, we haven't made much money. We have, however, paid all of the help and expenses and purchased equipment. This year we should make a profit.

It is hard to establish yourself in a new market. When your customers get to know you, they will come back. It is pretty tough to sell your product the first time. The second time it is easier. Having a number of varieties ripening at different times is very helpful because your selling season will be longer.

Direct Marketing on the Farm: Roadside Stands and U-Pick Operations

U-Picks. At U-pick farms, consumers harvest the produce themselves. Instead of paying packers, shippers, and brokers to market their crops, on-farm marketers sell directly to consumers at their farms. Some farmers sell all of their crops this way, while others compliment it with other marketing methods. Most U-picks are operated by fruit and berry growers but there are some vegetable operations. Some U-pick farms include variations such as rent-a-tree.

Rent-a-tree. An individual or family rents a tree (or it could be a row of strawberries or tomatoes) for one season. The farmer does all the work in the field until harvest, then notifies the renter that the produce is ready. The renter harvests the tree—all at one time or during several visits. A rental contract should state the rights, privileges, and restrictions of the renter, and let the renter know that fruit quantity and quality are not guaranteed. Some farmers do guarantee a minimum amount of fruit. Farming risks are assumed by the renter.

Roadside stands. Roadside stands range from seasonal wooden stalls to year-around rural attractions which include pie shops, gift boutiques, and refrigerated produce displays. Some roadside stands specialize in a single product, such as strawberries or apples. Others offer a wide variety of fresh and processed fruits, nuts, and vegetables.

When deciding whether to set up a roadside stand, consider:

- **Laws and regulations** (local, state, federal):

  Signs and parking—Check with the highway department regarding parking and entrance requirements and outdoor advertising signs. Check with your County Planning Commission for information on sign construction.

  Health—Check with local environmental or public health departments.

  Zoning and buildings—Contact your County Planning Commission. Laws usually permit growers to construct and operate roadside markets on their farms. Building permits may be required. The California Uniform Retail Food Facilities Law (CURFFL) exempts direct marketing farmers from most structural and operational requirements if the market is on land controlled by the farmer and the only items sold at the market are eggs, fruits, nuts, and/or vegetables. You may only sell your own products. CURFFL is enforced by county environmental health offices. Interpretations of CURFFL may vary from county to county.

  Taxes—Check state and federal laws.

  Employees—Check state and federal rules.

  You do not need a certificate from the Agricultural Commissioner.

■ **Insurance.** Make sure your insurance covers U-pick or a roadside stand. Most insurance companies in California will not write new policies for U-pick operations involving ladders.

■ **Location.** Poor location is one of the major causes of failure. The ideal location for a roadside stand is on a straight, level stretch of land adjacent to a major roadway. (To obtain an estimate of the number of motorists who drive by your farm, call the California Department of Transportation.) The entrance should be close enough to the road so it can be seen by passing motorists, but not so close that entering and exiting the parking lot is a hazard. Most people will only drive 10 to 25 miles to a farm. If your farm is not located within a 25-mile radius of an urban population or tourist attraction, establishing a profitable on-farm direct marketing business could be difficult, although some rural markets have done well with much of their business being repeat customers. To increase your potential for success, consider joining a farm trail organization or developing your farm into a rural attraction. Neighboring roadside stands or U-picks are not always a source of competition. You may grow a crop that compliments crops grown on surrounding farms. You can even be successful if several farmers in a particular area grow the same crop. Consumers are more likely to visit a farm if they can also stop at other farms while they are in the area.

■ **Attracting customers.** Attracting customers from the road requires attractive signs, easy access from the highway, an attractive entrance, and adequate parking, including parking

Arrowhead Farm, a Pick-Your-Own Operation
By Howard W. (Bud) Kerr, Jr.
USDA, CSRS, Office for Small-Scale Agriculture, Washington, DC

We purchased 17 acres in 1970. Looking back at how we operated Arrowhead Farm, a pick-your-own (PYO) farm, enables me to convey to you my first-hand experience with direct marketing.

Many people see agriculture as a "growing" business; however, the growing is only part of the puzzle. Besides the growing, issues such as financing, management and marketing should be paramount.

Our farm business was set up as a part-time business. My full-time job provides the necessary capital for my small farming venture.

We grow peaches (14 different varieties to bracket the season), thornless blackberries and strawberries. The business hours are from 12 noon until 5:30 p.m., Tuesday, Thursday, Saturday and Sunday. No one is allowed to pick earlier or later with the exception of one special group. The one hour prior to opening is for the handicapped only. This controls early birds. We've received excellent comments about this practice.

We are located on a small hard surface road very close to Baltimore. If there was a problem, it was too many customers coming to the farm if a newspaper advertisement was used. Early on we tried to establish a unique customer following and ultimately dispensed with newspaper advertising.

Three words describe the ideal direct farm market outlet: location, location, location. However, we find other words have considerable impact on repeat business: neat, neat, neat! Fresh paint on buildings, clipped green grass, flowers and landscaping, safe and secure parking areas and clear signs are appreciated. Instead of words, put pictures on signs wherever possible; a drawing of a large red strawberry whets appetites and brings cars right to the field.

for motor homes or other large vehicles. Attracting customers can be done through word of mouth (which follows a quality product), membership in a farm trails organization, and newspaper and magazine advertisements.

■ **Hours of operation.** Large on-farm operations are typically open eight to 10 hours a day, seven days a week during peak season. Small stands are often only open on weekends, when traffic is highest.

■ **Building design.** Roadside stands need easy, well-

Once people are trained to your system, there are very few problems. Remember, you want repeat customers.

We use a variety of selling techniques to enhance the business and our image. Here are some that we employed and others I think worth trying:

Be interested in your customers' families. Get to know their names, and those of their children. Ask where they work, what they do for a living.

Participate in community activities: church groups, garden clubs, service organizations, women's groups, health clubs, any place customers may be found. We learned what people were thinking and doing, their interests, etc.

Try some special events, like demonstrating how to prune fruit trees or how to plant strawberries. Once we had a local TV station out to film two ballerinas dancing to the music of a cello player in the middle of our beautiful in-bloom peach orchard. The customers gained were more affluent than the average—we didn't have to worry about bad checks!

We have a contest called "Pic of the Week." Our farm patrons submit photographs of members of their family at our farm. All pictures become our property and weekly we select the winner. Their prize is a flat of berries or basket of peaches—whatever is in season. We make sure different people win the prize each week. After each marketing season, we mount all winners' pictures in a frame for display. Over the years, the winners keep coming back to see how they looked a few years ago.

As the people weigh their strawberries and check out, we gain additional sales with special items. We often sell "Cackle Berries"—12 large strawberries in a white egg carton. It's a special gift for someone special—the nurse at the old age home, the neighbor shut-in, etc. We get a premium price for those berries which we pick prior to opening.

Always service the immediate fruit needs of your customers; however, never miss the opportunity of reminding your customers about the next fruit available. Once we all wore T-shirts, during strawberry season, that said "Reach for a Peach."

Provide a shady comfortable spot for older customers to rest. We always keep available "Wipettes" and a first aid kit. Have your employees trained in CPR—you never know when this skill will be needed.

Be very sensitive to nature, conservation, and environment. The perimeter of our 17 acres is posted with bluebird nesting boxes. Also, a martin box adorns the highest point in our peach orchard. For many years, a red fox raised a family in a hillside den just below the peach orchard and people loved to watch them from far off via telescope. And, there was another advantage to hosting our resident red fox. Never did we experience damage to a fruit tree being girdled by mice or rabbits.

You can increase PYO sales if your field employees have available extra picking flats and cheerfully suggest to pickers that their filled containers be taken to cool and retrieved when the customer is ready to check out.

A business that markets directly to the customer does not develop completely in a single season. The farm image—neatness, cleanliness, the logo, signs, personnel, quality of fruit, printed material and access roads—are all extremely important. City customers especially enjoy the farm outing and recreational aspects. However, they may be critical of muddy or dusty roads, inadequate parking, poor sign directions, weedy fields, or rude personnel.

Since accidents may occur when crowds come to the farm, be sure to obtain adequate insurance coverage. The normal farm insurance policy most likely will not cover accident liability or even the buildings used in the business.

defined access from the parking area to the store entrance; neat, attractive displays; clean walls and floors; attractive packaging; clear, legible signs stating size and price of products, and well-defined discount or special display areas. Place products according to size and popularity. There should be adequate space between aisles (four or five feet), proper placement of and adequate counter space for cash registers, accurate and easy-to-use cash registers and scales.

- **Customer relations.** Be prepared to answer questions about crop varieties, recipes, and growing methods. Become acquainted with your customers on an individual basis. You must enjoy having people come onto your property. You have to keep smiling even if branches are broken, kids throw tomatoes, cars back into your fence, or people try to steal. The customer is always right! Sales people should be easily identifiable, perhaps with special aprons or shirts, knowledgeable about the product and its uses. They need to pay personal attention to the customers. Public relations—advertising, signs, good humor, sincerity, and interpersonal relations—are part of the success formula that develops between producer and customer.

- **Labor.** Most small farmers rely on family members for labor.

Remember: Several years may be needed to build the business to an acceptable level.

Farm Trails

Farm trails are organizations of growers in a specific region. They publish a map indicating where their farms are and when consumers can buy produce. Farm trail organizations may also advertise on the radio and in newspapers. Some have annual harvest festivals. Growers usually pay a membership fee to cover the costs of producing the map, advertising, and other activities.

Farm trail organizations in California (there were about 20 in 1992) are listed in the California Agricultural Directory. See the list of additional resources and references at the end of the chapter.

Selling to Restaurants

There are both advantages and disadvantages to selling to restaurants. The advantages are:

- You have a consistent market.

- Your income may be higher than if you sell through a broker, though not as high as through farmers' markets or selling directly from your farm.

- You have personal contact with the users of your products (the chefs) so you get feedback about taste, whether it is the right product, packaging, if the size is right.

- You can test new products. If you have a new product that is successful in restaurants, it likely will be successful with the general consumer.

■ You can get a good reputation for your farm. Your customers will discuss your farm and products with other chefs, which may generate more business for you. Sometimes a restaurant will print your farm name on the menu. With a good reputation your farm name will become recognizable, like a brand name, which helps increase the price.

There are also big disadvantages to selling to restaurants.

■ It takes a lot of time: telephoning, visiting chefs, delivering two or three times a week, keeping the books, making sure the customers are satisfied. If you make deliveries to major cities, you may get into time-consuming traffic nightmares!

■ It is a low volume business. If you sell at a terminal market, you can sell a truck load of produce. When you sell to restaurants, you sell a few handtrucks full.

■ Chefs are artists and sometimes hard to work with.

If you want to sell to restaurants, do your homework. Define your target market. Is it white tablecloth restaurants with entrees costing over $12 a plate? Is it large volume restaurants where they buy large quantities or seconds? Find restaurant reviews in the newspapers. Get a book that rates restaurants. Get the names, addresses, and phone numbers of restaurants. Get as much information about the chefs and the restaurant as you can before you approach them.

Send the chefs a letter telling them who you are and that you want to make an appointment to sell your product. When you visit, bring samples. Chefs like to see you—as long as you don't call at 6 p.m. or noon! Telephone or visit (with an appointment) between 2 and 4 p.m.

Be prepared for your visit. Have a price list, a crop list (what you think you will have, the volume, your delivery schedule). You need a business card. You need to be able to explain your payment policy. If you are going to offer credit (which is recommended) you need to explain your terms. You need to tell them what your minimum order will be. They like to cut, taste, and look at the food. Invite the chef to your farm.

Chef at Sacramento Summer Harvest Tasting

Don't try to serve too many restaurants. Start slowly and build up the business.

It is much better to have too few customers that you are serving well than too many that you are serving badly. Chefs talk to each other and if you miss a delivery to one restaurant, the other chefs are going to hear about it and your business can tumble.

You need good communications with the chefs. Find out what products they want, including size and packaging. Serve them as closely as you can without being their servant or employee. If you try to pick and pack differently for each chef, you'll exhaust yourself. Strike a balance between serving their needs specifically and coming up with general solutions that work with all your customers.

Don't just deal with the executive chef. Deal with at least two decision makers in the restaurant because chefs come and go frequently. If you only talk to one person and that person leaves, you could lose their business overnight.

Remember: the chefs don't really need you. They have larger suppliers. But, they buy directly from growers because they like having a real farmer visit, explain growing practices, possible uses for products, harvest schedules, etc.

It is easy to get caught up in the excitement of making deliveries and wheeling and dealing. You can lose track of how much money you are actually making! Figure out how many customers you can effectively serve and what order size you need to be profitable. If the business builds slowly, decide how long you will give the customer to get up to the order size you need, and be willing to stop deliveries or charge for delivery if the order size isn't sufficient.

Finally, calculate the costs and benefits of using a food service purveyor. Sometimes it makes sense to use a middleman. They perform valuable services that you may be able to use.

Mail Order and Catalog Sales

If you sell products through the mail or catalogs, your customer base will be almost unlimited, but you will have high costs for promotion, packaging, shipping, and postage. You have to have a high value product to justify the costs. You have to have a product that can be sent through the mail without damage. Dried and processed goods are obvious possibilities and, with the advent of overnight and express mail, even some fresh items can be sold through mail order.

You'll need to develop and maintain a customer mailing list. If you already sell your products directly, at a roadside stand for example, get names and addresses from a guest book or from customer receipts. These customers are already interested in your product. Ask customers for names of people they think might be interested in your product. Get names and addresses from your friends. Send copies of your brochure to friends to give to their friends. Hand out your brochures at your roadside stand or farmers' market booth. You will build your list over time.

You can also buy or trade customer lists. But before you do, find out about your potential customers—age, education, family size, age of children, occupation, income, recreational activities they enjoy, political affiliation, etc. You don't want to waste a lot of money sending fancy catalogs to people who aren't interested in your product. If you already sell your product directly, give your customers a survey that asks such questions. You can use the results for developing mailing lists.

A Chef's Perspective

By Mark Casale,
Chef at Dos Coyotes, Sacramento

There are three things I look for when you come to sell me produce:

1. Quality—your products must be of superior quality.

2. Cost—your products must be priced competitively.

3. Accessibility—you must be able to provide me with what I need when I need it.

Superior quality can mean different things. It can mean you grow the best beefsteak tomato in the world and sell it at a reasonable price. It can mean you have a product that is superior because it is unique. I am always looking for different products to experiment with and present to my customers. If you offer me a purple and red striped tomato, I will probably buy it.

You have to sell at competitive prices. Restaurants have large budgetary constraints. We don't make a lot of money. People think, "Wow, it's packed; they're making tons of money." We don't. We make between one and 10 percent net profit. We make 10 percent in a big month like December, one percent in a slow month like January. Some months we lose money. Don't expect a high price.

Organic growers' costs are higher but don't expect to triple the price; I can't afford it. I can't afford to use all organically grown produce. If I'm putting your organically grown zucchini into my minestrone soup that has non-organically grown onions in it, it really doesn't make any difference whether it is organic or not. Organically grown is a tough sell, except in the lettuce market where people think uncooked foods are more healthy if they are organically grown.

You have to be consistent with your price. I realize you are very seasonal, but if you are going to sell me basil for the summer I want the same price all summer. I have to set my menu price. If I buy tomatoes in the winter at $40 a flat and in the summer at $5 a flat, I can adjust my menu seasonally. But, during a single season, because my menu price isn't going to change, I don't want your prices to change.

The next thing is delivery—can you get it to my restaurant when I need it in the amount I need. This is where I failed with several growers. I use 12 to 15 growers throughout the year, which is a lot of people to deal with. Some of them are good, some aren't so good. Some have gone by the wayside because they couldn't get me the products when I needed them. I had one person delivering beautiful organically grown lemons, oranges, and avocados from Southern California. He'd come once every two weeks. I'd buy 15 cases at a time. But he'd show up at 6 p.m. on Friday nights—my busiest time. He wasn't able to change his time so the relationship fell apart.

You have to let me know how much of a product I can expect. If you are a very small grower and you have just a little bit of a certain crop, you have to tell me, "This is all I have—three cases. There isn't going to be any more. Use it for something special this weekend." I might buy it, especially if it is something unique that I can't usually get. But you have to tell me I can't expect it every weekend. If you have a lot of a product and you want to set up a consistent relationship with me, that's great too. Let me know, "This is how much I have over this period of time; I'll be able to deliver to you 15 pounds of mushrooms every two weeks at this cost." That is something we can work with.

Consistency is the most important concept I want to relate to you. I need to know what you are selling me: It has to be a consistent product. It has to be delivered consistently. It has to have a consistent price. In the restaurant business we strive for consistency. If you're not going to be able to deliver consistency to me I'm not going to be able to work with you.

Soon a relationship will develop between us. You come in once a week and we talk and become friends and trade planting ideas and I give you recipes and you give me little starter plants for my home garden. Then we can expand our business. Maybe you have too much of a given product and you have to get rid of it. You might telephone me: "Mark, if I cut the price a little bit, will you take this?" That's the sort of relationship we're going to develop. I enjoy those relationships. You're trying to sell to me. I'm trying to buy from you. We're both involved in presenting food. As long as it works out for both of us and we're both making money and the public is pleased, it should be just fine.

You can buy lists from mailing list companies. They can select names by geographic area, income, age range, etc. For each refinement of the list you pay more. Response rates to direct mail are typically one to three percent. The more accurately targeted the list, the better the response rate. A list with 10,000 names might cost $350 for a one-time use. Once someone responds to the mailing, you can add the name to your list and contact that person as often as you wish.

To promote sales initially you will need a catalog or brochure. It can range from a simple one-page brochure to a multi-page, full color catalog. It should reflect the image you wish to project and it should appeal to your targeted audience.

Advertising in magazines and periodicals is another way to advertise your product. Magazines and newspapers already have a targeted audience. Classified ads in newspapers and circulars are inexpensive and don't require design and art work.

Keep track of your responses from direct mail and advertisements. Find out which promotions and publications were most effective. Calculate what it costs to reach 1,000 people. These records also help you determine who your customers are.

Make it easy for your clients to order your product. Use postage paid return envelopes. Have a phone number that can be used 24 hours a day (turn on an answering machine when you can't answer the phone), and take credit cards. Consider getting a toll free 800 number. Some people won't bother to write a check and mail a form but will phone in and order using a credit card.

After your initial sale, you need to promote further sales—an important part of a successful mail order business. Be sure to include a new catalog or re-ordering information.

The packaging and presentation of your product is important. Use good, strong shipping boxes. It is less expensive to use standard sizes for your packages than to have them custom made. Consider having an attractive label designed for your product. Gift packs are popular for the holidays and other special occasions. In the package you send a customer, include recipes or free samples.

Determine postal costs and delivery times. Check to see if sales tax is required. Give yourself enough time to fill your orders. During the holidays expect a lot of orders at the last minute. Be sure you can do what you say you will.

Keep the business simple, particularly when you are starting out, and have a quality product.

Community Supported Agriculture (CSA)

Community supported agriculture, community sustained agriculture, and subscription farming are all names for a group of

CSA Resources and References

Bio-Dynamic Farming and Gardening Association, Inc., P.O. Box 550, Kimberton, PA 19442. They sell a 10-page booklet called Community Related Agriculture, rent and/or sell a video called It's Not Just about Vegetables, and try to maintain a list of all CSAs in the country. They are one of the original organizations working with CSAs in the US.

Cohn, Gerry (editor), 1994. Community Supported Agriculture Conference. Proceedings of a conference at the University of California, Davis, December 6, 1993. Davis, CA: UC Small Farm Center. 51 p.

Groh, M. and S. McFadden. 1990. Farms of Tomorrow: Community Supported Farms, Farm Supported Communities. Kimberton, PA: Bio-Dynamic Farming and Gardening Association, Inc. 140 p.

Van En, Robyn. 1988. Basic Formula to Create Community Supported Agriculture, Great Barrington, MA: Indian Line Farm. 60 p.

customers who pay ahead of time for a season or a year and receive food on a regular basis. The farmer and consumer share in the risk of farming and in the rewards.

Customers buy a share, which costs a set amount per season or year. This provides the farmer with cash for farming and a reliable marketing outlet. The shareholders get a variety of locally grown, fresh produce. Generally the farmer makes weekly deliveries to one or more central locations where the shareholders pick up their food.

A CSA gives people a connection with the source of their food. They pay the true cost for the food, including supporting and building the farm.

The farm needs to grow a large mix of crops to provide the shareholders with a variety of goods, year-around. Early in the year, they may receive lettuces, spinach, peas, green onions, squashes, and other vegetables. As the season progresses, carrots, potatoes, broccoli, and fruits are delivered, followed by sweet corn,

Decater Farm
by David Visher, Small Farm Center

Steve and Gloria Decater's bio-dynamic farm is one of a rapidly growing number of farms around the world practicing Community Supported Agriculture (CSA).

They have farmed in Covelo, an isolated part of eastern Mendocino County, for over 15 years. The 40-acre farm provides most of the vegetables and some of the meat and eggs for about 100 families. About three acres are planted with vegetables and 20 acres are used for hay crops. Other land is used for animals, an orchard, berries, buildings, and gardens.

A fundamental difference between this farm and nearly every other is that the nearly 100 families that the farm feeds have a personal stake in the farm's health and productivity.

Steve explained, "We are trying to recreate a conscious connection between the consumer and the producer. The marketing system blocks people from coming into a connection with the food system. For people coming together in a community to take care of the environment, the food is a byproduct."

There are about 55 families who collect their produce weekly at an organic broker in San Francisco. Seventeen more families receive their food at Willits, an hour and a half away, and the remaining 30 families come to the farm.

Steve and Gloria maintain the cohesion of this community through a newsletter and planned visits by the members. "We have field days at least twice a year. People can come to work on the farm or just come to get a sense of life here. We like having that connection with them so they can think of the farm as their place."

Gloria said, "There are several interlocking communities in the CSA. There is the membership community, the local farm community, and the community on the farm. We try to keep the on-farm group healthy with work meetings and community meetings where concerns are aired. It is getting stronger and more effective. We are learning how to make it work. We are excited about it even if it is pretty intense.

"Our motivation is not money, it is to create the farm as a whole farm and to learn in that experience. We are living really close together and we have found that, if we are not united in spirit, it is impossible to do the physical work of the farm."

tomatoes, peppers, and melons. Farms located near metropolitan areas tend to be more successful in subscription operations than those located in rural areas.

Additional Resources and References

California Agriculture Directory, including Oregon and Washington. 1992. Sacramento, CA: California Service Agency. Addresses and phone numbers for farm cooperatives, federal and state marketing orders and commissions, farm trail groups, county agricultural commissioners, certified farmers' markets, and more.

Directory: Information Sources for Marketing California Fresh Fruits and Vegetables. 1990. Pub. 21480. Oakland, CA: ANR Publications. 32 p.

Facilities for Roadside Markets. 1992. Pub. NRAES 52. Ithaca, NY: Cooperative Extension. 32 p.

Gibson, Eric. 1993. Sell What You Sow! The Grower's Guide to Successful Produce Marketing. Carmichael, CA: New World Publishing. 304 p.

Jolly, Desmond. 1987. Regulations Governing Contracts between Growers and Handlers of Agricultural Produce: A Primer for Small-Scale Producers. Pub. 21425. Oakland, CA: ANR Publications. 8 p.

Luebbermann, Mimi. 1993. Pay Dirt: How to Raise and Sell Herbs and Produce for Serious Cash. Rocklin. CA: Prima Publishing. 209 p.

Produce Handling for Direct Marketing. 1992. Pub. NRAES 51. Ithaca, NY: Cooperative Extension. 26 p.

Stone, Bob. 1988. Successful Direct Marketing Methods. Lincolnwood, IL: NTC Business Books. 575 p.

Wampler, Ralph L. and James E. Motes. 1984. Pick Your Own Farming: Cash Crops for Small Acreages. Norman, OK: University of Oklahoma Press. 194 p.

VI.
ENTERPRISE IDEAS

Introduction

Every farmer has to decide what to produce or raise. The selection of enterprises determines if you will make money and whether your other farming goals will be met.

You need to determine your farming goals (see Chapter 1 and worksheets) and inventory the resource requirements of the enterprises you are considering (see Chapter 2 and worksheets at the end of the book), and study the markets (Chapter 4 and 5 and the worksheets).

You need to:

- Develop a list of alternative enterprises.

- Select your enterprises.

- Analyze compatibility among enterprises.

This approach is appropriate for farmers who are already engaged in farming as well as beginning farmers. Repeat the exercise periodically. Opportunities as well as goals change over time.

Chapter authors: Patricia Allen, former Coordinator, Small Farm Center, UC Davis, now Administrative Analyst, Agroecology Program, UC Santa Cruz; Karen Klonsky, Extension Specialist, Agricultural Economics Department, UC Davis

Keep written records on each step of the process. This forces you to give clear answers and is useful for making decisions in the future.

Develop a List of Alternative Enterprises

■ Which enterprises are predominant in your area?

■ Are you considering enterprises that have potential in your area but have not yet been established?

■ What crops or livestock have been raised on your land in the past?

■ At which enterprises are you most successful: livestock, field crops, orchard crops, small fruits, vegetables, ornamentals, growing transplants, raising seed?

Select Enterprises Compatible with Your Resources

Evaluate the potential for each enterprise on your list by using the appropriate worksheets at the end of the book. Compare resources needed to resources available. This will require a good deal of homework. Talk to other growers about their experiences with the enterprises you are considering. Check with your farm advisor. Don't forget your public library and local college campus specialists and their written materials.

Analyze Compatibility among Enterprises

Before making final decisions, consider the relationships among enterprises. You may have enough labor to produce an enterprise only if you don't select another labor intensive enterprise.

When resources are required can be as critical as the amount of resources required. A monthly chart of resource needs for each enterprise will be helpful.

Advantages to having several enterprises on a farm are:

■ There is less risk. The expenses of production failure and/or poor prices are spread over several commodities. Your cash flow and profit will be less variable in a diversified operation.

■ Resources are used more evenly throughout the year. This is helpful if you rely on hired labor because you can offer work for a longer time.

■ A range of products increases your access to markets. A buyer is often more interested in a grower who can supply a number of commodities so he or she does not have to buy from many different growers.

■ Crop rotation and crop mixing are effective methods of pest control and increasing soil fertility. These production practices—including inter-cropping, cover crops and green manure crops—may help control pests and improve soil fertility.

■ Combining livestock production with fruit and vegetable production helps with weed control and improves soil fertility through addition of manure.

Enterprise Possibilities

Livestock
Bees (honey, pollinators, queen rearing, etc.)
Beneficial insects
Birds (exotics, game birds)
Bull frogs
Chickens (and eggs for eating or hatching)
Ducks (rare breeds)
Earthworms
Fish (food fish, bait, pets, fingerlings for stocking)
Geese
Goats (milk, meat, wool)
Llamas
Ostriches and emus
Pigeons (squab)
Rabbits (meat, pelts, pets)
Sheep (meat, wool)

Forestry
Christmas trees
Fuel wood trees
Specialty lumber trees
Windbreak trees

Tree Fruits/Nuts
Apples (less common varieties)
Apricots
Avocados
Citrus (less common varieties)
Dwarf fruit trees
Figs (dried, fresh)
Peaches (less common varieties)
Pears (Asian)
Persimmons
Pistachios
Pomegranates

Small Fruits
Blueberries
Cane berries (blackberries, raspberries, etc.)
Grapes
Kiwis
Strawberries

Vegetable and Herb Crops
Amaranth
Arugula
Asparagus
Beans (fresh)
Broccoli
Cabbage (special varieties)
Cactus
Carrots (special varieties)
Cauliflower
Celeriac
Chard
Chayote
Collards
Corn (sweet, Indian, baby)
Cucumbers
Dandelion greens
Eggplant
Endive and chicory (raddichio, frissee)
Fennel
Garlic (regular, elephant, braids)
Herbs
Jerusalem artichokes
Jicama
Kale (colored)
Leeks (baby)
Lettuce (leaf, bib, baby, gourmet)
Melons (special varieties)
Mustard greens (Japanese)
Okra
Onions (green, special varieties)

Oriental vegetables
Parsley
Parsnips
Peas (sugar, black-eyed)
Peppers (bell, chili, unusual varieties)
Pumpkins
Radishes
Salsify
Shallots
Squash (summer, winter, baby, blossoms)
Taro
Tomatillos
Tomatoes (fresh, cherry)
Turnips
Watermelons (unusual varieties)

Ornamental and
Nursery Crops
Cut flowers (especially varieties not grown
  by large scale growers)

Dried ornamentals for arrangements
Nursery stock (especially rare varieties)
Seasonal ornamentals (e.g., holly, Indian corn)
Seedlings (ornamentals, vegetables)
Turf

Other Crops
Fiber crops
Guayule
Hay (small bales)
Mushrooms (exotic varieties)
Oil crops (sunflower, safflower, jojoba)
Seed (commercial, rare varieties)

Other Ideas
Custom machine operation
Farm management services
Specialized services
Vacation farms
Trucking for neighbors

Additional Resources and References

Farming Alternatives, a Guide to Evaluating the Feasibility of New Farm Based
Enterprises. 1988. Pub. NRAES-32. Ithaca, NY: Cooperative Extension.

Innovative Rural Enterprises. 1991. Pub. IES 153/ESO1897. Columbus, OH: Ohio
State University. Lists of references on alternative agriculture enterprises, from
alpaca to wood drying.

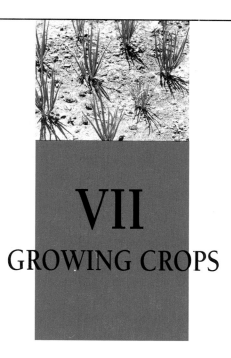

VII
GROWING CROPS

Water Management

Water Needs

Crop water needs vary with season, climate, soil type, plant type and growth stage, and rooting depth. However, you can estimate your water needs.

In general, the absolute minimum, continuous-flow water requirement from a well is five gallons per minute per acre, or 7,200 gallons per day, for drip irrigation. It takes about 7,000 gallons of water to irrigate an acre of mature fruit trees or vegetables every day in the heat of the summer.

To estimate water needs, determine your soil type (see page 77), consult your Cooperative Extension farm advisor, check with nearby farmers who produce similar commodities, and read about the water needs of your crops or livestock. Tables on water needs for most crops in most regions of California are available from your farm advisor.

Two other considerations in determining water needs are rooting depth and

Chapter Authors: Aziz Baameur, Farm Advisor, Riverside County Cooperative Extension; Shirley Humphrey, Staff Research Associate, Small Farm Center, UC Davis; Pedro Ilic, Farm Advisor, Fresno County Cooperative Extension; Faustino Muñoz, Farm Advisor, San Diego County Cooperative Extension; Eric Mussen, Extension Apiculturist, Entomology Extension, UC Davis; Richard Smith, Farm Advisor, San Benito, Monterey, and Santa Cruz Counties Cooperative Extension; Michael W. Stimmann, Statewide Pesticide Coordinator, Environmental Toxicology, UC Davis; Paul Vossen, Farm Advisor, Sonoma County Cooperative Extension

How to Estimate Water Needs

To estimate water needs, select the peak-use period. For example, assume you will grow five acres of watermelons. Through research you have found that during the peak season this crop uses about 0.25 inches of water per day.

0.25 inches per day x 5 acres = 1.25 acre-inches of water per day.

If your irrigation system is designed to run 10 hours per day:

1.25 acre-inches ÷ 10 hours = 0.125 acre-inch per hour.

One acre-inch = 27,154 gallons.

0.125 acre-inches per hour x 27,154 gallons per acre-inch= 3,394 gallons per hour.

How many gallons per minute is this?

3,394 gallons per 60 minutes = 57 gallons per minute (gpm).

Adjusted for 80 percent efficiency of the irrigation system, you will need 20 percent more water or,

57 gpm ÷ 0.8 =71 gpm.

You need a flow of 71 gallons per minute. You can irrigate one half of the acreage each day. This would allow you to use half as much water on any given day, or alternatively you can double the time of operation (if your system has been designed to do so). If the operating time is doubled, the required flow can be halved; in this example, to 35 gpm.

evapotranspiration. The rooting depth of a crop determines how much moisture depletion it can tolerate before it becomes stressed. Evapotranspiration is the amount of water lost from soil and plants because of evaporation and transpiration.

Plant water requirements are given in acre-inches as well as gallons. The chart "How to Estimate Water Needs" will help you convert the figures to the measurement you use most often. An acre-inch is the number of gallons needed to cover one acre one inch deep with water (27,154 gallons).

Water Needs and Climate
Total water requirements vary with climate:

- Plants in marine areas (fog-influenced climate where temperatures rarely exceed 75° to 80° F during the summer) use about 12 to 16 acre-inches of water per growing season.

Gallons of Water Vegetables Need Daily in Different California Climates

Climatic Conditions	Gallons Needed Per Day Per Square Foot	Gallons Needed for One Acre of Vegetables or Mature Orchard with a Cover Crop
Early spring-late fall; Foggy, cool day; Evapotranspiration—.10 inches per day	.062	2,715
Spring or fall; Some fog, warm day; Evapotranspiration—.20 inches per day	.125	5,431
Mid-summer; No fog, hot day; Evapotranspiration—.25 inches per day	.156	6,788
Mid-summer; Very hot, 100° F, windy, Evapotranspiration—.30 inches per day	.187	8,146

■ Plants in coastal cool areas (intermediate climate zone with some fog and a few hot summer days) use about 16 to 26 acre-inches of water per growing season.

■ Plants in coastal warm areas (intermediate climate zone with little summer fog and several hot sunny days) use about 26 to 36 acre-inches of water per growing season.

■ Plants in warm valley areas (interior valley climate with many summer days exceeding 90° F) use about 36 to 48 acre-inches of water per growing season.

Irrigation Systems and Management

Choose an irrigation system—surface, sprinkler, drip or trickle—that provides enough water but doesn't cost too much to install and operate.

Surface Irrigation

Any irrigation that uses the soil surface to deliver water is called surface irrigation. Depending on the crop, region, topography, and water cost, you can use basin, border, furrow irrigation, or a combination. For row crops, including vegetables and cotton, furrow irrigation is used most often. Basin and border irrigation are usually used for tree and grain crops.

In basin irrigation, water is ponded in small basins. There is little or no runoff, though on some low-rate infiltration soils, water may be drained off after enough water has percolated into the soil. Basin irrigation is simple and inexpensive.

In border irrigation, 30- to 100-foot wide levees are formed along the borders of the fields. The fields are graded so they slope. Water is discharged from a ditch or pipe along the upper end of the field and flows across the field to the lower end, where the runoff flows into another ditch.

Furrow irrigation is similar to border irrigation except that water flows through ditches that are six- to 12-inches wide and three- to six-inches deep.

Surface irrigation requires less equipment expense than sprinkler or drip irrigation, but can use more water if runoff is not recovered and reused.

Sprinkler Irrigation

In sprinkler irrigation, water moves through pipes to the sprinkler heads or nozzles. It is ejected into the air and falls to the ground at varied distances. Many growers use sprinklers to germinate a crop before switching to a different irrigation form. With tree crops some growers start with basin irrigation, change to a spitter sprinkler, and then to a revolving sprinkler. Sometimes, only revolving sprinklers are used, particularly where water costs are not high.

Your choice of crops will help you determine whether to install permanent sprinklers, portable sprinklers, or sprinklers installed on short movable hoses attached to permanent risers.

Permanent sprinkler and drip systems are convenient and often reduce labor requirements. On large orchards, portable sprinklers are most efficient since labor is

Water and Irrigation References

Determining Daily Reference Evapotranspiration. 1992. Pub. 21426. Oakland, CA: ANR Publications. 12 p. How to calculate evapotranspiration for various California locations.

Using Reference Evapotranspiration and Crop Co-efficients to Estimate Crop Evapotranspiration: Agronomic Crops, Grasses and Vegetable Crops. 1987. Pub. 21427. Oakland, CA: ANR Publications. 12 p. Use with Pub. 21426.

Using Reference Evapotranspiration and Crop Co-efficients to Estimate Crop Evapotranspiration: Trees and Vines. 1987. Pub. 21428. Oakland, CA: ANR Publications. 8 p. Use with Pub. 21426.

Drip Irrigation Management. 1981. Pub. 21259. Oakland, CA: ANR Publications. 44 p. System design and operation for efficient use of water and other drip-injected materials.

Drought Irrigation Strategies for Deciduous Orchards. 1989, Pub. 21453. Oakland, CA: ANR Publications. 16 p. How to limit water loss and time unavoidable plant water deficits.

Drought Tips for Vegetable and Field Crop Production. 1989. Pub. 21466. Oakland, CA: ANR Publications. 24 p. Strategies for conserving irrigation water.

Irrigation Scheduling: A Guide for Efficient On-Farm Water Management. 1989. Pub. 21454. Oakland, CA: ANR Publications. 80 p. How to measure soil, atmospheric, crop, and irrigation factors to determine the most efficient irrigation schedule. Includes historically averaged evapotranspiration figures for 208 California locations.

Questions and Answers about Tensiometers. 1981. Pub. 2264. Oakland, CA: ANR Publications. 12 p. Soil and water status in the root zones to use as a guide to regulating irrigation practices. A tensiometer is a closed tube filled with water that is inserted into the ground and shows how much water is in the soil.

Water-Holding Characteristics of California Soils. 1989, Pub. 21463. Oakland, CA: ANR Publications. 92 p. Charts arranged by county to show ability of various California soils to store and supply water as needed.

involved in checking sprinklers even when they are permanently installed. Under these conditions, it is little more expensive to move portable sprinklers than it is to check the permanent installation.

The water pattern should cover the root zone and, to avoid runoff, the water should not be applied faster than the soil can absorb it. This is particularly true on soils with high clay content or on compacted soils. Excessive water may cause other problems, such as root rot, so a system offering localized control features (individual shut-offs) may be desirable for particularly sensitive crops.

There are seven commonly used sprinkler systems that vary in capability and appropriateness: portable sprinklers, side roll systems, traveling gun systems, center pivot systems, linear move systems, low energy precision application systems, and solid set systems.

A sprinkler system is a good substitute for a surface system where soils are not deep or too variable, or where topography is unsuitable. However, sprinkling water on some crops, such as cucurbits and strawberries, may increase disease problems. Salt in sprinkling water can damage fruit tree leaves.

Energy use, water pressure, labor, and management vary with each system.

Drip Irrigation and Micro Sprinklers

Drip irrigation is a low volume irrigation system designed to deliver water at a slow rate. The system can use any type of micro sprinkler (mostly for fruit trees) or drip irrigation (mostly for vegetables) except in extreme cases where soils are very fine or very coarse. Drip irrigation is suitable to most situations, even on steep slopes.

Drip irrigation is the frequent, slow application of water to soil through emitters that are located at selected points along water lines or on short branch leaders at each tree. The amount of soil wetted by drip irrigation is much less than that wetted by other irrigation methods. For newly planted trees, only 10 percent of the soil in the root zone may be wetted. Observations show that at least 33 percent of the soil in the root zone under mature crops should be wetted and that crop performance improves as the amount wetted increases to 60 percent or more. The amount of soil wetted depends on soil characteristics, irrigation running time, and the number of emitters used. Newly planted trees may need only one emitter, while larger trees could use eight or more.

(Large trees are sometimes watered with two mini-sprinklers that disperse the water over a larger root zone area.)

Drip systems require the frequent application of water to dilute salts and move them away from plant root zones, and to supply the daily water needs of plants on a low volume basis. One big drawback to drip irrigation is the initial cost of the system and the disposal of discarded drip tape. The system uses less energy and requires less pressure and labor than surface and sprinkler systems. However, it requires just as much if not more management skill.

Fertilizers that are highly soluble can be injected into drip irrigation water to avoid the labor needed for ground application. Greater control over fertilizer placement and timing through drip irrigation may lead to more efficient fertilizations.

Problems in drip irrigation most often involve clogging, poor uniformity of water delivery, salt accumulation, and runoff. Algal growth in the drip line may clog or cause expensive damage to elaborate drip systems. However, all these problems are easily solved with careful management. Additional problems with drip tape include: animals and insects that chew drip tape; water runs to the lowest spot—the ground must be reasonably level; pressure reducers may be necessary for some pumping systems; and drip tape is easily ruptured during mechanical weeding processes.

Water Conversion Factors

Area	1 acre = 43,560 square feet
Concentration	One part per million (1 ppm) = 1/1,000,000 = 1 milligram/liter
Flow rate	1 cubic foot/second = 449 gallons per minute or 26,940 gallons/hour
	1 gallon/minute = 0.01 acre-inch/4-1/2 hours
Volume	1 acre-foot of water = 12 inches over 1 acre field = 325,851 gallons
	1 acre-inch of water = 27,154 gallons
	1 cubic foot of water = 7.48 gallons

Soil Management

Soil provides physical support for plants and acts as a reservoir for water, air and nutrients. Soil texture—the mixture of different sizes of minerals in the soil—is coarse, medium, fine (also called sand, silt or clay, respectively), or a combination (loam). The finer the texture, the higher the water retention, so the interval between water applications can be longer. Soil high in clay is sticky, hard to dig or plow, dense, and its surface bakes in hot dry sunny weather. Sand is inert and water can rush through it and remove soluble materials. Silty soils have a slower water intake rate than sandy soils but their water holding capacity is higher. A clay soil can be lightened by adding sand, and a sandy soil can be made heavier by adding clay. Most soils are loams which are combinations of sand, silt, and clay. Loams are easier to work, more porous, and less likely to bake than clay, and retain moisture and plant food better than sand.

Soil is also affected by the amount of organic matter in it and tillage and cultivation practices. Soils should contain 25 percent air, 25 percent water, and 50 percent minerals and organic materials.

Study your soil's texture, structure, fertility, and, especially, the amount of time it takes to drain. Soils that don't drain well can be formed for better drainage. In deep, well-drained soils, trees don't get root rot and they produce more. The more care you take managing your soil, the better your crops will be.

Plants require a number of chemical elements. They get carbon from the air and the other nutrients from the soil surrounding the roots. A fertile soil provides these nutrients. Fertility can be improved by adding appropriate chemicals.

Plants need nitrogen, phosphorus, and potassium in relatively large quantities. Plants need smaller amounts of sulfur, calcium, magnesium, iron, zinc, manganese, copper, chlorine, boron and molybdenum.

Soil that has an appropriate amount of organic matter is more fertile and produces better crops. Organic matter can be incorporated into soil by plowing under cover crops or crop residues or by adding organic soil amendments and fertilizers. Manure and compost are commonly used to improve soil fertility. Kelp, rock phosphate, and worm castings are also used as soil amendments. Rotating complementary crops (legumes to grains to row crops) increases soil fertility. Crop rotation also reduces insect and disease problems, and maintains soil tilth.

Soil Nutrients
The key to good plant health is a balanced diet and regular feeding with an appropriate amount of nutrients. Excess nutrients, though they may not be obvious in the short run, can cause significant long-term problems.

Plants use most nutrients in the ionic or elemental form: nitrate (NO_3^-), ammonium (NH_4^+), potassium (K^+), phosphate ($H_2PO_4^-$), and other elements, not complex molecules (proteins, carbohydrates, vitamins).

Nitrogen
Nitrogen is used by plants as nitrate and ammonium. Nitrate is very soluble, but ammonium is rapidly converted to nitrate in well-aerated soils. Nitrogen is stored in soil organic matter (living, dead, and composted plants). Some nitrogen sticks to soil particles, but once it becomes soluble, it can be lost into the water. If soil dries out, nitrogen can be lost into the air as a gas. Thus, organic matter is important in nitrogen management because it allows the nitrogen to be released slowly, providing a continual supply for plants and serving as a buffer against leaching. In the absence

of large quantities of organic matter, water management and nitrogen management (including the method of application, timing, and amount) become even more critical.

Phosphorus
Since phosphorus is not very soluble in water, it moves little from where it exists

Factors Influencing Nutrient Needs

Aeration	Roots require oxygen for respiration and nutrient uptake.
Light	Nutrient uptake is a result of photosynthesis; it cannot happen without light. Also, some crops are day-length sensitive.
Moisture	Roots don't grow to water, they grow in water. Nitrogen moves with water, both toward and away from plant roots.
Microorganisms	Because soil microbes break down organic matter into usable forms, they are important to soil fertility. Many form a symbiotic relationship with plants. However, they may also compete with plants for nutrients.
Nutrient concentration	Increased nutrient concentrations in the soil solution or on the soil particle surface result in increased nutrient uptake.
Organic matter	Organic matter provides a reservoir for nutrients over a long period of time and buffers against leaching.
Pest and disease presence	Pests and diseases prevent plants from absorbing nutrients efficiently.
Plant age	The older the plant, the less efficient it is in absorbing nutrients.
Root system	Most vegetable crop root systems are shallow. Crops with extensive root systems are more efficient "gatherers" of nutrients.
Soil pH	Nutrients vary with the type of soil. Some nutrients (phosphorus, iron, manganese) are more available in more acid soils. Others are more available in more alkaline soils (calcium, magnesium, molybdenum, sulfur). Soils with pH lower than 6.0 or higher than 8.0 may need additions of lime, gypsum, or sulfur.
Soil type	Soil is classified as sand, silt or clay. Clay soils have a greater capacity to store nutrients than sandy soils.
Temperature	Crop roots and tops grow faster at higher temperatures, which increases their nutrient needs. Roots absorb phosphorus poorly at low temperatures, thus more is needed close to the roots in cool or cold soils.

Nutrient Needs

	Heavy users	Moderate users	Low users
Nitrogen	Broccoli	Cucumber	Asparagus
	Cabbage	Lettuce	Bean
	Carrot	Pepper	
	Cauliflower	Tomato	Filbert
	Onion	Summer squash	Grape
	Potato	Sweet potato	Kiwi
	Sweet corn		Olive
	Winter squash	Apricot	Persimmon
		Apple	Pomegranate
	Almond	Cantaloupe	Quince
	Berries	Cherry	
	Citrus	Chestnut	
	Nectarine	Fig	
	Peach	Pear	
	Walnut	Pistachio	
		Plum	
		Prune	
		Watermelon	
Phosphorus	Beet	Bean	
	Broccoli	Cantaloupe	
	Cabbage	Cucumber	
	Carrot	Onion	
	Cauliflower	Pepper	
	Celery	Potato	
	Lettuce	Summer squash	
		Sweet corn	
		Tomato	
		Watermelon	
		All fruit tree crops	
Potassium	Beet		
	Carrot		
	Celery		
	Potato		
	Winter squash		
	Berries		
	Grape		
	Prune		
	Walnut		

naturally in the soil or where it is added as fertilizer or is incorporated as part of organic matter. Phosphorus can be leached from the root zones into groundwater in very sandy soil. Generally, plant roots must grow to the phosphorus rather than the reverse. Phosphorus should be applied close to roots. Phosphorus deficiency is rare in tree crops because the extensive root systems reach the element.

Potassium

Many vegetables are heavy users of potassium. Many tree crops require the addition of potassium. Potassium is water soluble but does not move with the water because it sticks to soil particles. Except in very sandy soils, leaching is usually minimal. Most California soils are naturally moderate to high in potassium.

Soil Sampling and Testing

Have your soil tested to determine the amounts of various elements it contains as well as determining the pH and electrical conductivity of the soil. Soil tests can reveal nutrient deficiencies and point out areas high in nutrients. Most tests show soil pH, amounts of nitrates, ammonium, potassium, phosphate, and minor nutrients.

Since fields vary in soil types, many samples are needed from each field. One sample consisting of 20 cores or slides of soil for each 15 to 20 acres is usually necessary. Areas having different soil types or different fertilizers or other additions are sampled separately. Avoid unusual "patches" when gathering samples. Sampling soil by dumping a few handfuls of soil in a bag is a waste of time. Take your samples after disking the previous crop, but before you fertilize your next crop. Gather samples after irrigation.

Soil Management References

California Commercial Laboratories Providing Agricultural Testing. 1991. Pub. 3024. Oakland, CA: ANR Publications. 22 p.

Generalized Soil Map of California. 1988. Pub. 3327. Oakland, CA: ANR Publications.

Organic Soil Amendments and Fertilizers. 1993. Pub. 21505. Oakland, CA: ANR Publications. 36 p.
A thorough introduction to composts, manures and other organic soil materials, where you can get them, how they work, and what they can do for your farm.

You will need a clean plastic bucket or plastic bag, spade or soil probe. If you don't have a soil probe, use a spade. Take one-half inch thick slices then trim the sides leaving a one inch strip. Sample eight inches deep. Put these cores into the bucket or plastic bag, break the clods and mix the soil thoroughly. There will be about one quart of this well mixed soil for each sample you send to the laboratory.

Label the sample with your name, address, and sample number. Write down the origin of the sample, past crops, quantity and types of fertilizer used on previous crops, herbicides used, and nematode or disease problems.

The laboratory you select to process your soil samples should use University of California test methods.

Excerpted from Otto, H.W., Roy Branson, Kent Tyler. 1983. Guide for Fertilizing Vegetables. Pub. ANRP012. Oakland, CA: ANR Publications.

Pest Management

Throughout history, damaging pests have been a problem for farmers, causing damage to seeds, seedlings, mature crops, and stored products. Pest management has always demanded a significant amount of a farmer's time and effort. At times pests have devastated agricultural production and caused famines which starved millions of people.

Examples of pests are insects, parasites, weeds, nematodes, worms, fungi, snails, slugs, rats, mice, deer, gophers, rabbits and diseases. (Any unwanted plant, animal, or microscopic organism can be considered a pest.) Many different pests can affect almost every plant or animal. Sometimes the effect is beneficial or causes little or no damage. You should be aware of potential problems, however, because many can be controlled by prevention.

Insect and disease management manuals for many crops are full of information on life cycles, spray schedules, avoidance techniques, and registered pesticides (organic and conventional).

Certain pests are major problems every year, such as powdery mildew on grapes, brown rot on stone fruits, flea beetles on summer crucifers (broccoli, cabbage, cauliflower, etc.), or codling moth on apples and pears. These pests can destroy 100 percent of the crop under certain conditions. If you can, choose a site that is less likely to experience devastating pest problems such as well drained isolated locations with good air movement, and select varieties and rootstocks that are resistant to pests.

Noxious perennial weeds can only be eliminated before planting. These include: nutsedge, umbrella sedge, Bermuda grass, Johnsongrass, bindweed, wild blackberry, poison oak, and sheep sorrel.

Microbial diseases such as viruses, bacteria, and fungi are in the environment and may infest plants when conditions are right. Most soils contain weed seeds that can remain dormant for years until appropriate conditions exist for their germination. Arthropods such as insects, mites, and millipedes can fly, be blown, crawl, or be transported into crops. Growers should be aware of as many of the various cultural, biological, and chemical ways of dealing with these problems as possible.

An agricultural chemical may be used only for crops and pests for which toxicological research has resulted in official "registration" with both the Environmental Protection Agency and the state regulatory agency. Most chemicals are registered for major crops grown in substantial quantities throughout a region. Because research is expensive, registration frequently is lacking for minor crops.

Integrated Pest Management
Integrated pest management (IPM) combines several techniques for dealing with pest problems. IPM may reduce the need for pesticides by incorporating various practices, such as crop rotations, scouting to see what pests are current, weather

monitoring, use of resistant cultivars, timing of planting, mating disruption with pheromones, and biological control of pests.

Biological control is the use of natural enemies to control pests. Although biological control agents occur naturally in the environment, they often are purchased for augmentative release into crops for pest control. This involves releasing increased numbers of already occurring predators and parasites at the most vulnerable period in the host life cycle. There are many commercial insectaries providing biological control agents, and the use of this approach is expanding every year. To ensure success, users of commercially available predators and parasites need to understand the ecology of the environment, including the target pest population, and the predator or parasite to be released.

One of the essential components of IPM is knowing the pest. The effective use of IPM depends on knowing the life cycle, ecology, and physiology of the pest in order to intercept it at its most vulnerable stage. The pest species and population densities must be monitored closely to make decisions on when and how to proceed. Preserving beneficial organisms is important, and IPM relies on substitutes for conventional pesticides like traps and barriers, pesticides of low toxicity such as soaps, oils and microbials, and changing planting, irrigation or cultivating procedures.

Some Important Terms

Pests are organisms that attack, compete with, or adversely affect crops, humans, homes, pets, and livestock. Examples are: insects, weeds, nematodes, parasites, fungi, snails, slugs, rats, mice, and animal and plant diseases. Any unwanted plant, animal, or microscopic organism may be a pest.

Pesticides are substances used to control pests. Included are: insecticides, herbicides, defoliants, fungicides, nematicides, and rodenticides. Even a common substance like water or salt may be a pesticide if used to control a pest.

Pesticide Use Regulations

California has the most stringent pesticide regulations in the nation. Every farmer who uses pesticides is required to know, understand, and follow the pesticides laws and regulations. There are federal, state, and county pesticide regulations in effect in California.

Instructions on pesticide labels must be followed. The label is a legally binding document. If you fail to obey the instructions, you are violating the law and may be subject to severe legal penalties. County Agricultural Commissioners enforce the program.

Major chemical companies produce, and "register" with government regulatory agencies, pesticides for major crops such as small grains, corn, soybeans, cotton, and cattle. Registering a pesticide for a minor crop is not economically justified for the companies. Therefore, it is often difficult to find effective pesticides registered for minor crops. To pesticide manufacturers, even almonds, tomatoes, lettuce, tree fruits, grapes, and citrus are minor crops! The federal government has established the Inter-Regional Project Number 4 (IR-4) to assist farmers with minor crop pesticide registration problems. There is an office at UC Davis.

Pest Management References

Grape Pest Management. 1992. Pub. 3343. Oakland, CA: ANR Publications. 412 p.
Details the least costly, most effective pest control measures for control of grape insects and diseases.

Pests of the Garden and Small Farm: A Grower's Guide to Using Less Pesticide. 1990. Pub. 3332. Oakland, CA: ANR Publications. 286 p.

UC IPM Pest Management Guidelines. 1990. Pub. 3339. Oakland, CA: ANR Publications. 900+ p.
Information on insect, mite, disease pests, nematodes, and weed pests for 27 major crops.

Integrated Pest Management Manuals. Oakland, CA: ANR Publications.
Comprehensive, practical guides on the latest pest information, illustrated with over 150 color photos and dozens of drawings.
   Alfalfa Hay, Pub. 3312, 98 p.
   Almonds, Pub. 3308, 152 p.
   Apples and Pears, Pub. 3340, 216 p.
   Citrus, Pub. 3303, 144 p.
   Cole Crops and Lettuce, Pub. 3307, 112 p.
   Cotton, Pub. 3305, 144 p.
   Potatoes, Pub. 3316, 146 p.
   Rice, Pub. 3280, 108 p.
   Small Grains, Pub. 3333, 126 p.
   Tomatoes, Pub. 3274, 104 p.
   Walnuts, Pub. 3270, 96 p.

Suppliers of Beneficial Organisms in North America. 1992. Sacramento, CA: California Environmental Protection Agency.

All pesticides registered in the US are either for general or restricted use. General use pesticides are available to anyone. They are sold in supermarkets and hardware stores, and they present minimal risk to the user or the environment when label directions are followed. Permits are not required, but agricultural uses must be reported to the County Agricultural Commissioner monthly.

Restricted use pesticides are more hazardous and not available to the general public because they present health or environmental hazards. They are available only to holders of pesticide use permits. To be granted a pesticide use permit you must pass a state examination and obtain the permit from your County Agricultural Commissioner. Permits may include conditions and restrictions beyond those on pesticide labels, such as potential environmental impacts to groundwater, wildlife, and people.

Everyone who uses pesticides for agricultural uses must receive training first. The County Agricultural Commissioner requires the farmer-trainer to keep a written record of training given.

If label instructions are followed, adverse environmental impacts are minimized. The less that gets into the environment the less there is to cause disruptions. All farmers should review their pest management programs and adhere to IPM principles.

For additional information, contact your County Agricultural Commissioner, your local farm advisor, pesticide companies, pesticide dealers, licensed pest control advisors, and certified pesticide applicators.

Cover Crops

Cover crops are grown between plantings of marketable crops for adding organic matter to soil, and protecting against erosion. They have been used in California agriculture for more than 50 years. Examples of cover crops grown in California include vetches, clovers and cereals. Most cover crops are legumes, because they incorporate nitrogen into the soil, though non-legumes are used in some situations. They can be grown summer or winter; however, most California growers use winter species. Cover crops offer a practical means of supplying organic matter to soil to keep it in a high state of productivity. The decaying organic matter can provide nitrogen and other soil nutrients for succeeding crops. Cover crops also provide habitats for beneficial insects, improve soil tilth, facilitate water penetration, and increase the diversity of micro flora in the subsoil.

Planting cover crops may result in some negative impacts as well. They may attract pests, require additional seed, irrigation water, and labor. Fall cover crop plantings prevent the ground from being worked up to allow for spring planting flexibility. In some areas, like the semi-arid regions of the Central Valley, it may not be practical to use cover crops because they may require substantial amounts of water for irrigation. Also, revenue-producing acreage is reduced when a cover crop is grown. However, the cost of planting and maintaining a legume cover crop can be thought of as the cost of producing nitrogen and improving soil quality.

Good growth of a well-inoculated legume (legume cover crop seed is inoculated with a nitrogen-fixing bacteria) will frequently add 100 to 200 pounds of nitrogen to an acre of soil. Cover crops grown for their nitrogen are often called green manure. The amount of organic matter produced, the quality of the organic matter, the ease of incorporation, and the amount of residue left in the soil may affect subsequent cultural operations. These are governed by the type of cover crop chosen, the inoculation used, the planting date, and when the cover crop is disked in.

For more information read Covercrops for California Agriculture, 1989, Pub. 21471, Oakland, CA: ANR Publications, 24 p.

Crop Diversification

Farmers who manage a few acres often have highly diversified operations. During a single growing season they raise a variety of vegetables and fruits, and sometimes poultry and livestock.

Diversification enhances economic stability by spreading risks over a greater number of crops. Ideally, the crop mix should be complementary; that is, all practices should be performed without competition for labor, equipment, or management expertise. This is not always possible because of factors which are beyond control, such as unusual weather conditions or pest infestations. Growing a mix of crops can help prevent damaging organisms from becoming numerous enough to cause significant crop damage. While one crop may be destroyed under such circumstances, other crops will remain unaffected.

Crop Rotation

Crop rotation is the successive planting of different crops (usually two to five different crops) in the same field. Growers rotate a variety of crops, often including a cover, or green manure, crop. A successful rotation maintains fertility with a minimum of fertilizer inputs, controls pest and disease problems, reduces weeds, reduces soil erosion, and provides adequate income. Crop rotation recycles nutrients, breaks pest cycles, and helps maintain a balance between soil organic matter accumulation and decomposition. Crop rotation implies rotation of tillage and other cultural practices.

A number of different rotation schemes are feasible. An example of a typical rotation for a small vegetable farmer is:

Year 1—Cherry tomatoes

Year 2—Onions

Year 3—Squash

Year 4—Beans

If you stagger your plantings, you can harvest longer.

Production Costs and Returns

The cost to grow fruit ranges from $5,000 to $15,000 per acre before the first fruit is picked. For vegetable crops, the cost ranges from $700 to $2,500. Vegetable costs can go up to $4,500 per acre if season extenders such as plastic tunnels and mulches are used. For field crops typical costs per acre are $150 to $400 before harvest. These costs do not include land, interest payments, taxes or depreciation. Overhead and time costs are lower for vegetable farms because of a quicker (three- to six-month) return on investment compared to two to six years for fruits.

Many Cooperative Extension farm advisors provide cost analyses for the major crops produced in their counties. These analyses include costs of materials, cultivation practices, and labor. Often they calculate estimated income depending on yield per acre. If you are considering growing non-traditional crops, a farm advisor may be able to give you an enterprise budget for crops that are comparable to the crop you are growing.

Many factors affect your anticipated yield and the amount you can charge for your products (location, weather, insects, weeds, diseases, farming skills, variety). The selling price will also depend on your buyer, when you sell (prices fluctuate daily), supply on the market, quality of your product, and your marketing knowledge and skills.

Growing Vegetables

Growing vegetables requires year around planning and scheduling. For example:

In the spring you plant, fertilize, irrigate, weed, and thin, all the while keeping careful records to help manage next year's crop. You prepare for harvest and arrange for extra labor, boxes, cold storage, and distribution.

In the summer you complete spring jobs while the crops mature, and begin to harvest.

In the fall you continue to harvest and market your goods, clean the fields, plant cover and winter crops.

Enterprise Budget—Bell Peppers
Fresno County, 1990

Pre-harvest costs	Per Acre	Per Box
Land preparation		
Labor (3 hours @ $5.50/hour), tractor, implements	$ 50.25	
Fumigation		
Metham sodium (40 gallons @ $5.00/gallon)	200.00	
Application sprinkler rental	345.00	
Labor (5 hours)	27.50	
Tilling: labor (3 hours), tractor	34.50	
Drip tape for irrigation	245.00	
Fertilization		
Pre-plant fertilizer	68.60	
Labor (1.5 hours), tractor	10.75	
Nitrogen (200 pounds)	84.48	
Plants: 14,500 @ $65/1,000	942.00	
Planting: labor (8 hours), tractor (4 hours), planter	68.00	
Irrigation electricity (4 months @ $50/month)	200.00	
Pest control		
Material	40.00	
Labor (4 hours), tractor	46.50	
Plant removal		
Shredder rental	11.25	
Labor (1 hour), tractor	11.50	
Miscellaneous	100.00	
Office expenses (6% of pre-harvest costs)	149.00	
Total pre-harvest cost	**2,634.33**	**$3.29**
Harvest costs		
Pick, haul and pack	1,400.00	
Total harvest cost	**1,400.00**	**$1.75**
Depreciation		
Tractor and equipment	155.00	
Irrigation system, mains, filter, etc.	50.00	
Truck, trailers, shed. etc.	73.00	
Total depreciation	**278.00**	**$0.35**
Interest on investment at 8%		
Half cost of $15,000.00 for 10 acres	60.00	
Tractor and equipment, 14% for 10 years	108.50	
Total interest on investment	**168.50**	**$0.21**
TOTAL COST OF PRODUCTION	**$4,480.83**	**$5.60**

Notes:
- Potential yield: 800 30-pound boxes per acre.
- Labor is included. If you provide labor, expenses will be reduced.
- The beds are 3 feet apart with two rows 1 foot apart; plants are 2 feet apart in rows.
- Bell pepper is one of several crops on a 10-acre farm.

In the winter, with fewer crops growing and less daylight hours to work outdoors, you plan for the coming seasons; buy supplies; clean and repair tools and equipment; plan major building projects (such as a new workshop or an improved irrigation system). This is the time to go over your records; decide whether to plant the same thing or something different; plan test plots, crop combinations, planting schedules, and crop rotations. Prepare greenhouse and hardening off space for starting transplants. Order seeds and plants. Determine weed and insect control procedures. Schedule fertilization, irrigation, weeding, and thinning. Sign up new buyers.

Raised Beds

Permanent raised beds are three to five feet wide and four to 12 inches deep. A significant amount of organic matter is worked into the beds to make a very fertile and well structured soil which may support plant spacing that is much closer together than in a conventional system.

The major advantage of this system is more yield per square foot because of the closely spaced plants. The closely spaced plants are better able to compete with weeds. Usually raised bed systems are designed to be worked by hand. There is little foot traffic on the beds, and the soil does not become compacted.

Labor and management are more intensive than in conventional row crop production.

Conventional Planting Beds

Annual bed systems are used by farmers who use machinery rather than hand labor and rotate their crops. They grow the crop, harvest it, and plow everything under at the end of the season.

Growing Fruits and Nuts

In an established orchard, you need to prune and thin, deal with insects, diseases, and weeds, fertilize, irrigate and finally harvest.

Pruning is done mainly in winter to balance vegetative vigor and good fruiting, but also in the summer to enhance fruit color and reduce vigorous tree growth. Pruning branches or canes achieves a balance between fruiting wood and vegetative wood that will fruit in the future. Because pruning is labor intensive, it is costly. It requires a great deal of knowledge and experience.

Enterprise Budget for Establishing a One-Acre Orchard

Water source

Power hookup	$ 1,000
Well	3,000
2 pressure tanks	1,000
Switch box and labor @ $10.00/hour	1,000
Total water source	**6,000**

Irrigation system

Main lines and faucets	150
Drip tubing	120
Emitters and micro-sprinklers	160
Valves, pressure regulators, filters	290
Sand separator	200
Control station	350
Miscellaneous	30
Labor @ $7.00/hour	700
Total irrigation system	**2,000**

Plants (small quantity prices)

Blueberries @ $3.50 each (24 plants)	84
Raspberries @ $0.75 each (50 plants)	38
Fruit trees @ $7.60 each (200 trees)	1,520
Labor @ $7.00/hour (51 hours)	357
Total plants	**2,000**
TOTAL ESTABLISHMENT COSTS	**$10,000**

Notes:
* Costs not included: Land, interest on investment, lost income from alternative employment, equipment (tractors, mower, sprayer), energy (fuel and electricity), taxes, management

It improves fruit size and reduces biennial (alternate year) fruiting. Thinning by hand is the most expensive operation in the orchard other than harvesting. Apples are the only chemically (hormone) thinned trees. Some trees require no thinning.

Spraying for pests must be done regularly. In the dormant season, growers spray for aphids and mite eggs, San Jose scale, and fungal infections. At blossom time, apples and pears are sprayed for scab, at three- to four-week intervals for codling moth (worms), as well as when aphids, mites or other pests arise. Stone fruits are sprayed at blossom for rot, in November for shothole fungus. Peaches are sprayed in February for peach leaf curl. Grapes require spraying or dusting every two weeks for powdery mildew. Several other sprays may be necessary as pest problems arise. Organic production may require more frequent spraying than conventional production since most allowable materials have low toxicity and low residual effect.

Controlling weeds to avoid competition with plants is essential (mulching, herbicide, flaming, or hoeing).

Fertilizing—applying nitrogen to young plants and micro nutrients to established plantings—must be scheduled according to recommended practices in the area. Compost preparation is an on-going process. Adequate nutrients are particularly important for young trees in the establishment year.

Irrigating—soil must neither be dry nor saturated. It should be kept moist all summer with approximately one inch of water per week and trees should never be allowed to stress during establishment (first five years).

Harvesting and handling fruit usually requires seasonal labor. To ensure an attractive product at the time of sale, you need a packing line, fruit bins, picking bags, boxes, fruit trays, and refrigeration. Rapid cooling and cold storage is expensive, but without it the quality of your fruit will be below standard and you cannot compete with other growers.

Rootstocks
For every fruit and nut variety and site there is usually one best rootstock. When you plant an Elberta peach tree, you are not actually growing Elberta peaches; you are growing Nemaguard or another rootstock with the Elberta variety grafted to it. The rootstock choice is the single most important biological influence on fruit and nut tree production. Get advice from local farmers and your Cooperative Extension farm advisor.

Hedgerows and Trellises
Fruit crops are commonly grown in hedgerows on size-controlled rootstocks. Plants can be pruned, thinned, sprayed, and harvested from the ground. Return on investment is much sooner, quality is better, and small trees are more economical to farm. Initial investment is higher—both to build a trellis system and to plant more trees per acre than are planted in traditional orchards. Many tree crops that used to be planted 2 feet by 20 feet apart on a five-year system are now planted at a spacing of 10 feet by 16 feet or even closer on dwarfing rootstocks.

Investigate the many types of trellis and pruning systems in use and choose one that fits your farming methods and pocketbook. Apples in California, for example, have shown little benefit from the extra expense and greater labor required to attaching trees to trellis systems. Wine grapes, however, can be greatly influenced by trellis system height and width for both production and quality.

Managing the Floor

A cover crop is often grown on the orchard floor. Choose a cover crop that will supplement fertility and help control erosion. Cover crops such as grasses, grains, and low growing early maturing annuals come back from seed every fall. These are usually controlled by mowing. Other cover crops are planted every fall and disked or tilled in the spring as green manure. Since cover crops compete for water with the main crop, drip irrigation or micro-sprinklers are normally used to irrigate the trees.

Growing Field Crops

Small farm operators grow field crops as part of crop rotation plans, as cover crops to add nitrogen to the soil, for human consumption, as animal food and bedding, for erosion control, and for spinning or milling. The most commonly grown field crops in California are wheat, barley, oats, rice, cotton, beans, peas, alfalfa, field corn, grain sorghum, and pasture grasses. Some farmers are growing and testing the market for little known field crops, such as amaranth, colored cottons, and unusual rices.

The Small Grains

The small grains are wheat, barley, and oats. Wheat is grown throughout California, principally for milling and animal feed. Barley is grown mainly for feed. Oats are grown for forage as well as human consumption.

Wheat, barley, and oats are cool season crops, planted to take advantage of rainfall. Varieties must be adapted to planting elevation. Barley does less well than oats or wheat on heavy, poorly drained soils, but is more tolerant of saline soils. Small grains can be grown on irrigated or non-irrigated farmland. A growing season total of 24-30 inches of water is sufficient.

The small grains are usually planted from November into early January. In the cooler, high mountain valleys, they can be planted in the spring when danger of severe frost is past. Seed is either broadcast or drilled into the soil. Nitrogen fertilizer is used before planting, depending on soil chemistry. Some soils also need phosphorous. Weeds are numerous unless controlled by use of weed-free seed and mowing. Diseases are best combated by the use of treated seed. Weeds, diseases and insects sometimes must be controlled chemically.

Grains are harvested in late spring or summer. In the cooler northern valleys, spring sown grain is harvested in August. Stored grain should be checked periodically for insect infestation.

Alfalfa

Alfalfa hay is grown in nearly every county in the state. It ranks high in both acreage and cash value. Alfalfa is grown for hay, seed and sprouts. It needs well drained, deep, medium-textured soils. A porous subsoil is essential for deep water penetration and root development. It is planted by broadcasting seed in both spring and fall. Inoculated seed is recommended if alfalfa is planted in fields not previously planted to alfalfa. Surface and sprinkler irrigation may be necessary in dry periods. This perennial crop can be harvested for three to five years.

Growing Specialty Products

As a small farmer you probably have limited resources, whether they be land, labor, financial, or other. But, you still want to make a living. The solution for many small farm operators is producing "specialty" products. If you decide to grow specialty crops, try something few or no other farmers are growing. Once there is plenty of a product, it is no longer so special. Growing a specialty crop requires experimentation both in the field and in the marketplace. Your local farm advisor may be able to give you information about a more common crop that is similar to the one you are testing. Library research and writing to universities or research centers located where the crop is commonly grown may help you gather information. You could be a pioneer in researching growing techniques and developing the market for a new product.

It is difficult to develop a market for a new item. There is considerable risk because market potential has not been tested. See Chapters 4 and 5 for marketing ideas.

Frieda Caplan of Frieda's Finest Produce Specialties in Los Angeles recommends you:

- Determine what can grow on your soils, given your climate and water availability.

- Find out what varieties of seeds and plants are available.

- Find out if there is a market for the product.

- Plant a small experimental plot or a few rows to get some experience.

Specialty products fall into three categories.

New, Unusual or Exotic Produce

Consumers like to try new varieties. A new color, flavor, size, or shape is appealing. A "new" item may be an unusual variety of a common crop such as lettuce. Lettuces are now available in numerous shapes, colors, and sizes with names such as red oakleaf, rouge d'hiver, and black-seeded simpson. Many specialty crops are of European, Asian or Latin American origin introduced to the US by immigrants. Examples are Belgian endive, bitter melon, radicchio, cilantro, daikon, and tomatillo.

Specialty Crop Resources and References

California Rare Fruit Growers, Inc.,
The Fullerton Arboretum,
California State University, Fullerton, CA 92634.
A non-profit organization that brings together people interested in producing rare fruits and publishes a bi-monthly magazine.

Kowalchik, Claire and William H. Hylton (ed.). 1987. Rodale's Illustrated Encyclopedia of Herbs. Emmaus, PA: Rodale Press. 545 p.

Morgan, Julia F. 1987. Fruits of Warm Climates. Winterville, NC: Creative Resource Systems Inc. 505 p.

Myers, Claudia, et al. 1991. Specialty and Minor Crops Handbook. Pub. 3346. Oakland, CA: ANR Publications. 37 p.

Stamets, Paul and J.S. Chilton. 1983. The Mushroom Cultivator. Olympia, WA: Agarikon Press. 415 p.

Stamets, Paul. 1993. Growing Gourmet & Medicinal Mushrooms. Berkeley, CA: Ten Speed Press. 552 p.

Stephens, James M. 1988. Manual of Minor Vegetables. Gainesville, FL: University of Florida. 123 p.

Whealy, Kent. 1989. Fruit, Berry and Nut Inventory. Decorah, IA: Seed Saver Publications. 366 p.
An inventory of nursery catalogs listing all fruit, berry and nut varieties available by mail order in the United States.

Whealy, Kent. 1992. Garden Seed Inventory. Decorah, IA: Seed Saver Publications. 502 p.
An inventory of seed catalogs listing all non-hybrid vegetable seeds available in the United States and Canada.

Yamaguchi, Mas. 1983. World Vegetables. Westport, CT: AVI Publishing Co. 415 p.

High Quality Produce

A product may be called "specialty" if it is grown and handled with exceptional care, such as tree ripened peaches shipped in special boxes and delivered within a few hours of harvest or organically grown strawberries when the market is full of conventional ones. To maintain such high quality requires hand labor and well thought-out packaging and delivery schedules.

Out-of-Season Fruits and Vegetables

Demand is always high for tomatoes in January or the first apple of the fall season. Maybe, through research or experimentation, you will discover a variety that matures much earlier or much later than what is currently available. You may want to construct greenhouses or use plastic row covers to extend the growing season.

Additional Resources And References

United Fresh Fruit and Vegetable Association, 727 North Washington Street, Alexandria, VA 22314.
They sell leaflets on more than 80 fresh fruits and vegetables offering data on produce growing, shipping and marketing, as well as historical and botanical information.

American Vegetable Grower, 37841 Euclid Ave., Willoughby, OH 44094.
Monthly magazine on culture, management, equipment, etc. for vegetable growers.

Harrington, Geri. 1984. Grow Your Own Chinese Vegetables. New York, NY: MacMillan Publishing. 268 p.

Kader, Adel (technical editor). 1992. Postharvest Technology of Horticultural Crops. Pub. 3311. Oakland, CA: ANR Publications. 296 p.

Lorenz, Oscar A. and Donald A. Maynard. 1980. Knotts Handbook for Vegetable Crops. New York, NY: John Wiley and Sons. 390 p.
A detailed reference on vegetable crops.

Sutter, Steve. 1993 (revised). New Federal Pesticide Safety Standard Issued. Fresno, CA: Cooperative Extension.

Sutter, Steve. 1993 (revised). Pesticide Safety Guide for Agricultural Workers. Fresno, CA: Cooperative Extension.

Ware, George, and McCollum, J. P. 1991. Producing Vegetable Crops, Danville, IL: Interstate Printers and Publishers. 607 p.

VIII
RAISING ANIMALS

Chickens, pigs, rabbits, goats, sheep, and sometimes horses and cows are typical small farm animals. These animals are complementary to the small farm—they eat crops produced on the farm or the crop residue, and their waste, mixed into the soil, recycles nutrients for plant use or acts as organic matter that improves soil structure. Your livestock enterprise can provide food for the family, supplement income, or provide a balanced operation. Choosing animals best suited to your needs and abilities will help determine whether you'll make a profit. Talk to others who are already raising the livestock you're considering. Listen to their histories and learn from them.

To make money, as a small farmer, with limited land, facilities, and finances, you may consider building a market for livestock that is not widely available, such as range chicken or miniature horses. Information may not be easy to find, but your Cooperative Extension farm advisor will give you information about similar breeds which will help you get started.

Basic Considerations

Legal Restrictions
Check local and legal restrictions regarding the types and numbers of animals that

Chapter authors: Fred Conte, Aquaculture Specialist, Department of Animal Science, UC Davis; Ralph Ernst, Extension Poultry Specialist, Department of Avian Sciences, UC Davis; John M. Harper, Livestock and Natural Resources Advisor, Mendocino County Cooperative Extension; Eric Mussen, Extension Apiculturist, Entomology Extension, UC Davis; Al Woodard, former Avian Scientist, Department of Avian Sciences, UC Davis may be kept on your property. There may be restrictions on the animals' proximity to dwellings. There are often rules on fences and structures for animals.

Other legal problems may occur after you establish a livestock operation. Sometimes neighbor dogs harass or kill animals. Sometimes neighbors complain about noise, dust, odor or flies. Many cases result in legal action.

The best way to familiarize yourself with local rules is to start collecting information from regulatory bureaus. Ask your local county Cooperative Extension farm advisor for suggestions. Contact county or city departments of planning and zoning, building, health, animal control, and the County Agricultural Commissioner. Check with state offices, including the State Brand Inspector at the Department of Food and Agriculture, and the Department of Fish and Game. These agencies are listed in the government section of your phone book. When you call, describe your situation and ask how your operation is affected by their regulations. Before you hang up, ask if they know of other departments you should contact.

Purchase Price and Where to Buy Animals

The purchase price of an animal varies with species, stage of maturity, health and condition, geographic location, sex, and whether it is a "grade," a purebred or a registered animal. Grade animals have parents of no particular breed. Purebred animals have parents of the same breed and often are registered with a breed association. Registered animals usually cost more than unregistered or grade animals because of the cost of maintaining specific traits of a particular breed and the market value of those traits.

Generally, the smaller the animal, the lower the price—you wouldn't expect to pay more for a rabbit than a steer. Younger animals and very old animals are sometimes less expensive because mortality is greater. Animals in poor health and condition are cheaper but obviously they are not recommended. Geographic location can influence price because of supply and demand.

Local or state livestock associations will provide names of people who raise livestock for resale. Registered breed associations have membership lists and will give you names of breeders in your area. Feed stores usually have public bulletin boards with advertisements of livestock for sale. Auction yards have sales at least weekly. Your farm advisor will know about many of these and can provide you with advice on local sources.

Feed Costs

Feed costs are usually the greatest variable expenses a livestock owner will incur. Your decision about what kind of animal to raise will be partially based on the volume of feed the animal will consume and your financial ability to provide it. The amount of food needed is related to body size. Requirements vary depending on the animal's stage of life. Different amounts of nutrients are required if the animal is immature, pregnant or lactating. The larger the animal, the larger its intake and the higher the cost to feed it.

Ruminant animals, those with multi–compartmented stomachs like goats, sheep

Be a Good Neighbor

by Aaron O. Nelson, Farm Advisor,
Cooperative Extension, Fresno County

Exotic animals and exotic animal owners are on trial. Most people view raising exotic animals as a new concept, a novelty, even a fad. Acceptance as a legitimate and permanent business must be established.

Fortunately, the novelty aspect of an exotics business provides a period of grace for you. People are not as likely to form prejudices, and will probably be initially supportive of a concept that piques their curiosity. But, because they will be watching your operation closely, the bad as well as the good will be accentuated.

How your neighbors accept this new component of agriculture is in your hands. You can create and maintain a good image, but you must work at it continuously. Lest you think that facts and logic alone are enough, consider the swine industry. Perception, not fact, causes them to be judged by different criteria than used for other farm animals. For example, some zoning laws allow commercial production of livestock species except for hog operations, which are required to have a conditional use permit. There is no objective biological basis for this, just a stereotypical image of pigs as filthy animals.

Don't let that happen to you and your business!

Thirty years of experience has taught me that disputes and conflicts with neighbors or governmental agencies usually revolve around a short list of factors. Concentrating on these will return big dividends.

Zoning—The rules are simple: raise them where you're supposed to; don't raise them where you shouldn't. The application is not so simple. Many counties do not have rules for where exotic livestock can be kept. Most counties equate an exotic species or breed to a corresponding domestic farm animal and apply the same zoning and density criteria. Zoning prescribes both what and how many you can raise. Conditional use permits may be granted, but you need to prepare a comprehensive plan to show your ability to deal with nuisance factors.

Manure—Most nuisance conflicts are precipitated by the accumulation of manure and its byproducts. Dealing with these problems beforehand is critical.

Odor—You may think it smells like money, but your neighbor will not be sympathetic. Removal and/or quick drying of manure and other organic wastes are the best solutions. Chemical treatment, masking odors, and other measures are less effective and more expensive.

Flies—Flies deposit their eggs in moist organic matter. During warm weather, you need to dry or liquefy manure. Liquefying will stop larva development and may help mechanize handling, but won't eliminate odor production. For most operators, drying is the answer.

Run-off—Problems with manure runoff are regulated by zoning laws and common sense. You'll need to take specific measures to prevent runoff, even during unusually heavy precipitation.

Groundwater pollution—Many agencies, especially the State Water Quality Board, are concerned about both chemicals and infectious organisms entering the groundwater supply. You can manage the manure from the equivalent of three to five dairy cows on an acre. Some soils and aquifer profiles may not support this high concentration. Consult an expert.

Dust—Whether from mineral soil or dry organic matter, dust is harmful to the health of your animals, you and your social relations. Wetting down pens and corrals may minimize this, but be careful not to create fly/odor problems. A good sod, or pasture, may be your best answer.

Noise—You may not be able to get your stock to take vows of silence, but you can locate them away from your neighbor's bedroom window.

Other pests and parasites—In addition to manure-bred pest insects, your animals may be a source of other pests such as mosquitoes, gnats, fleas, ticks and lice, which could spread to your neighbor's animals.

Solutions to many of these potential problems may require technical knowledge that you have not yet acquired. Consult experts such as veterinarians, university specialists and experienced producers. Read technical literature, available from your county Cooperative Extension office.

Visit your neighbors and explain who you are and what you will be doing. Invite them to visit. Ask for constructive criticism. Keep your operation neat, use paint and landscaping, construct visual barriers. You will find your efforts to be a good neighbor will also make your animals a welcome part of the community.

and cattle, and near–ruminants like rabbits or horses, can use low quality feed and they need only one or two types of feed. Fowl and hogs have more complex dietary requirements and need more complete rations. These higher quality diets cost more.

Using Pasture or Range

If you want your livestock to graze, you need to determine how much pasture or range the animals will need. The amount of land varies depending on species and life stage as well as soil type and topography, climate, and land and animal management practices. The problem of determining land requirements for different types of animals has been partially solved by the concept of animal unit equivalents (AUE).

Some Definitions

Animal unit (AU)

The equivalent of one mature (1,000-pound) cow that consumes an average of 2.5 percent of her body weight (25 pounds) of dry matter per day.

Animal unit month (AUM)

Potential air dried forage intake of one animal unit (1,000 pound animal) for 30 days—about 750 pounds.

2.5% x 1,000 lb. x 30 days = 750 pounds

Stocking rate

Amount of land allocated to each animal unit for the entire grazeable period of the year.

Approximate Forage Consumption

Animal	Type	Daily consumption (AUE)
Goats	Doe and kid pair	0.24
	Goat, mature, nonlactating	0.17
	Kid, weaned	0.14
Sheep	Ewe and lamb pair	0.30
	Sheep, mature, nonlactating	0.20
	Lamb, weaned	0.14
Horses	Horse, draft (mature)	1.50
	Horse, saddle (mature)	1.25
Cows	Bull (24 mo.+, 1,700 lb.)	1.50
	Bull (18–24 mo., 1,150 lb.)	1.15
	Cow and calf pair	1.35
	Cow, mature, nonlactating, 1,000 lb.	1.00
	Heifer, pregnant, nonlactating (18 mo.)	1.00
	Yearling (18–24 mo., 875 lb.)	0.90
	Yearling (15–18 mo., 750 lb.)	0.80
	Yearling (12–15 mo., 625 lb.)	0.70
	Calf (weaning to 12 mo., 500 lb.)	0.60
	Calf (weaning at 8 mo., 450 lb.)	0.50

From Valentine, 1990. Grazing Management. San Diego, CA: Academic Press, Inc. p. 276-293.

Once you know the forage consumption of an animal and have estimated forage production you can calculate how many animals your pasture or range can support and for how long. Rough estimates of forage production are available from the Soil Conservation Service or your Cooperative Extension farm advisor, or ask neighboring livestock producers (make sure their herds look healthy). But, your experience and monitoring of actual use will be your best information.

Climate, topography, and other factors complicate grazing capacity. The table below shows the effects of slope and canopy cover on estimated grazing capacity for three climatic areas in California.

Most pastures are greatly improved with good land and animal management practices. Fertilization, drainage, rotational grazing, regular clipping, weed control, pasture renovation, and reseeding improve pasture productivity. For example, a mature horse requires two and one-half to three and one-half acres of dryland pasture versus one acre of irrigated pasture.

To determine the number of AUs per acre per year that your range will carry, divide the value for your climatic zone, slope and canopy percentage by twelve. As an example:

The table shows on a Northern California range, with a canopy cover of 0 to 25 percent and a slope of under 10 percent, there are 3.5 AUM per acre available. Dividing by twelve results in about 0.3 AU per acre per year.

Estimated Grazing Capacity of Three Climatic Zones with Various Slopes and Canopy Cover

Area	Canopy Cover	AUM/acre by Slope Class			
		Under 10%	10–25%	25–40%	Over 40%
Southern California under 10" precipitation	0-25 %	0.7	0.4	0.3	0.1
	25-50 %	0.4	0.3	0.2	0.1
	50-75 %	0.2	0.1	0.0	0.0
	75-100 %	0.1	0.0	0.0	0.0
Central Coast, Central Valley Foothills, 10 - 40" precipitation	0-25 %	2.0	0.8	0.5	0.3
	25-50 %	1.5	0.6	0.4	0.2
	50-75 %	1.0	0.4	0.3	0.1
	75-100 %	0.5	0.2	0.2	0.1
Northern California over 40" precipitation	0-25 %	3.5	1.3	0.8	0.5
	25-50 %	2.8	1.0	0.6	0.3
	50-75 %	1.8	0.7	0.5	0.2
	75-100 %	0.9	0.3	0.2	0.1

From McDougald et al., Davis, CA: UC Davis Agronomy and Range Science, 1991. (The figures in this table are not for rotational grazing.)

3.5 AUM/acre ÷ 12 AUM/year = 0.3 AU per acre

How many acres are needed to support one AU?

0.3 AU per acre = about 4 (3.3 rounded up) acres for one AU

Once you have chosen the type of animal you wish to raise and have figured out how many of them your pasture can support, you can use several management tools for improving either the pasture's production or the harvest efficiency of your livestock. Pasture production (yield or quality) can be improved by fertilization, weed or brush control, irrigation and renovation and/or reseeding. Before undertaking range or pasture improvement, prepare a cost vs. benefit analysis based on recommendations from your local Cooperative Extension farm advisor.

Pastures cannot be grazed continuously without a period of rest for the plants to regenerate themselves or you will have overgrazing. Overgrazing, which kills plants, can lead to bare soil susceptible to erosion. Overgrazing is not a function of the number of animals on a pasture but how much time those animals are allowed to graze. Plants can be severely defoliated once and will recover if they are given time to regenerate and their nutrient requirements are adequate. Grazing animals are highly selective in their choice of which plants in a pasture they will eat even though avoided plants may be equally nutritious. This can cause overgrazing of a few plants in a pasture.

You can control overgrazing with a variety of management schemes which will result in increased harvest efficiency—more plants grazed to proper height without overgrazing a select few.

Time-controlled grazing (also known as intensive grazing management, high density-short duration, or mob stocking) divides the pasture area into multiple paddocks. The grazing animals are heavily stocked in a single pasture and then moved to a new paddock. The plant growth cycle is used to determine when animals are moved. This way more stock can be carried without overgrazing than under either continuous set stocking or other rest/rotational methods. The total carrying capacity as calculated from the soil type, canopy cover and topography for the pasture remains unchanged regardless of the management scheme. Under conventional set stocking or non-time controlled rest/rotational grazing schemes less than the total carrying capacity is stocked to prevent overgrazing.

Time-controlled grazing is complicated. For help in using it, consult your local farm advisor. If incorrectly managed, high density-short duration grazing can destroy a pasture very quickly.

Housing and Space Requirements

Just as humans require clean housing protected from the elements, so do animals. Like humans, adequate room to exercise is of paramount importance to the health and vigor of the animal. Size, age, sex, species and the number of animals will influence shelter and confinement requirements. A catalog listing USDA plans for farm buildings and other farm structures is available for reference in county Cooperative Extension offices. Plans may be ordered for $4 a sheet from Plan

Services, Agricultural Engineering Extension, University of California, Davis, CA 95616. Plans are also available from Midwest Plan Service, 122 Iowa State University, Ames, IA 50011-3080. Contact your local farm advisor for assistance once you decide on the type of livestock you wish to raise.

Guidelines in the chart are suggested minimum space requirements. Common sense should direct you to allocate enough space so your animals don't harm themselves. Too much space, on the other hand, can make catching an animal difficult and will certainly limit your potential production or profit. In warm climates or during the summer, plan on using the upper limits of space. Extra space will help the animals dissipate heat and remain cooler. This is especially important if you raise hogs.

Minimum Space Requirements for Livestock

Animal	Type	Floor Space (square feet)	Yard Space (square feet)
Poultry	Chickens, laying	2-3	
	Chickens, broilers	1	
	Ducks	3-4	4-5
Pigs	Sows before farrowing	15-20	15-20
	Sows with pigs, gilts	48	48
	Sows, mature with pigs	64	64
	Herd boars	15-20	15-20
	Swine, growing–finishing, under 75 lb.	5-6	6-8
	Hogs, growing–finishing, 75–125 lb.	6-7	7-9
	Hogs, growing–finishing, 125+ lb.	8-10	8-0
Rabbits	Rabbits	3-4	
Goats and sheep	Ewes, dry	16	16-20
	Goats, dry	16	200
	Ewes w/lambs, does w/kids	20	30
	Stud rams or bucks	20-30	30-60
Cows	Two years or older	40-50	300
	Cattle, yearling, finishing	30-40	125-200
	Calves, 350-500 lb.	20-30	100-175
	Cows, pregnant	100-120	1-2 acres
	Herd bulls	100-150	1-2 acres

From Ensminger, M.E. 1978. The Stockman's Handbook. (5th edition) Danville, IL: Interstate Printers and Publishers, Inc.

Health Care

Though you can administer routine health care yourself, you need a veterinarian to provide preventive care and to treat diseases, parasites, injuries and illnesses. Health care varies greatly between different species of livestock. Routine health care is more complicated and expensive for large animals. Your local veterinarian can give you a good idea of routine health care needed and an estimate of costs.

Managing

Knowing your physical limitations will help narrow your choice of what livestock to buy and make your project successful and enjoyable. The larger the animal the more physical strength is required. Managing a 600-pound steer is much more difficult than managing a 10-pound rabbit. Male animals tend to be more aggressive and larger.

You must be available to feed, protect, observe, and treat your animals several times every day. You'll spend time on record keeping, learning new information, and attending field days and meetings.

Marketing

Study auctions, cooperatives, and other buyers; learn what time of year to sell; decide whether you will slaughter before you sell. Where will you sell? Do you have transportation available? How much will transportation cost?

Poultry

Most poultry and egg production is on large specialized ranches. Egg collection, feeding and watering are usually mechanized. A few companies process most of the fryers for the commercial market. Turkey production is concentrated on a few large farms in the San Joaquin Valley.

A number of small-scale producers have been able to survive and compete in the specialty bird market. Most sell directly to customers. Carefully investigate the market for specialty poultry before starting production. A small distributor or processor may be looking for a certain volume of a specialty item; if more is produced, the price will quickly fall and so will your profit.

A few types of "specialty" poultry and "exotic" birds that small farmers successfully raise and market are chickens, ducks, quail, squab, turkeys, partridges, pheasant, and guinea fowl.

Chickens

Many small farmers can earn additional income by selling eggs directly to consumers or by producing eggs that command higher prices, such as brown shelled eggs, fertile eggs or free range eggs. None of these products are considered to be nutritionally superior by scientists, but some consumers are willing to pay a premium for them. Freshness, while not assured in these types of eggs, may be a factor in consumer decisions.

Recently, there has been an increasing market for dark feathered (usually red or black) meat chickens and White Silkie chickens. They are usually marketed live to small poultry markets or directly to consumers. Dark feathered meat-type stocks are available as day-old chicks. White Silkie chicks usually are produced on the ranch where they are grown. These chickens command premium prices and can be profitably produced on small farms.

Ducks and Geese

Most ducks grown in the United States are the pekin breed, which are used extensively for meat. The market for geese is limited, but they are popular on farms for eating weeds, as "watchdogs," and for home-produced meat and eggs. The duck raising business is dominated by a few very large producer-processors. There are several alternatives which are much more likely to be profitable for small farms.

The production of duck eggs (usually fertile) has proven to be a profitable enterprise on several California farms. Any breed can be used, but the most prolific layers are Khaki Campbell and Indian Runner. The eggs are often sold fresh, salted, and/or preincubated about 18 days. All of these are popular with California's large Southeast Asian population. The adult birds are often sold live at farmers' markets or to small processors where they bring a reasonable price for meat. Fertile duck eggs can usually be produced for about $1.00 per dozen in fixed and variable costs, excluding labor.

Muscovy ducks are raised for meat. They are often crossed with other domestic ducks (often pekin). The crosses are sterile and won't reproduce. The Muscovy has a much leaner carcass than other breeds of ducks and the meat sells for a higher price.

Some growers successfully market liver from heavily fed ducks for production of paté.

Japanese Quail

Japanese quail have been raised commercially in the US since about 1960. They were first imported for use as game birds but that venture

Poultry References and Resources

"Game Bird Letter" (newsletter). Davis, CA: Department of Avian Science, University of California.

Growing a Small Flock of Turkeys. 1981. Pub. 2733. Oakland, CA: ANR Publications. 10 p.
Selection of stock, housing and equipment, feeding and management, disease prevention, marketing and processing.

Raising Chukar Partridges. 1982. Pub. 21321. Oakland, CA: ANR Publications. 12 p.
Guide for commercial and novice breeders.

Raising Ducks in Small Flocks. 1977. Pub. 2980. Oakland, CA: ANR Publications. 11 p.
Information on breeds, housing, feeding, disease prevention, and marketing.

Raising Game Birds. 1978. Pub. 21046. Oakland, CA: ANR Publications. 24 p.
For beginners or experienced growers; includes information on pheasant, partridge, quail and wild turkey.

Raising Geese. 1975. Pub. 2225. Oakland, CA: ANR Publications. 8 p.
Breeds of geese and management practices.

"Squab Newsletter." Modesto, CA: Cooperative Extension.

Starting and Managing Small Poultry Units. 1975. Pub. 2656. Oakland, CA: ANR Publications. 17 p.
Includes information on selection of birds, feeding, housing, and health care.

Other sources of information:

California Department of Fish and Game
1416 Ninth St.
Sacramento, CA 95814.

Department of Avian Sciences
Cooperative Extension
University of California, Davis, CA 95616-8532.

A Small Poultry Farming Operation
by David Visher, Small Farm Center

Stuart Helfand raises poultry on five acres near Sacramento. He raises Silkie, Rhode Island Red, Black Australorp chickens, and squab. A Silkie is a black skinned bird that is considered by some to be an aphrodisiac. Helfand sells the birds alive directly to Bay Area Asian markets. By not slaughtering the birds, he simplifies marketing. If he or his family processed the birds without hired labor they would not be covered by inspection regulations.

"I let them catch their own birds; that way they have no complaints. They know what they want. I just bring up several birds at a time, and let them choose."

The road to success has not been without problems. One mistake he has not recovered from is turkeys. "They are still out there,

eating! I thought they would be profitable but they were just a hassle. The turkeys tied up a lot of room and money only to have people who had promised to buy cancel out at the last minute."

He has about 1,000 birds. "I wholesale Silkies for $4.50 apiece. They eat about 10 pounds of feed; that costs about $2.00. I have two customers who take a couple hundred Silkies a week. On two accounts that is 400 birds at $4.50 each. They also buy quail and red birds by the hundreds per week.

"I also work as a carpenter in order to put up more buildings without having to borrow money. The house payment is covered by my wife's salary. The poultry operation covers everything else. It is a good business."

was not very successful. Now they are sold for meat and egg production. The eggs are often marketed at farmers' markets and through specialty food stores. The key to success is careful planning and development of a reliable market.

The eggs sell for 8 to 12 cents each and live adults for meat sell for $.75 to $2.00 each. The price for meat birds depends on quality, size and demand. A jumbo strain which is about twice the size of the original imports (which weigh from 3.5 to 6 ounces) has been developed.

Quail production usually requires the production and incubation of fertile eggs, since day-old chicks are not widely available. The birds mature at five to six weeks of age and are prolific layers. About 12 ounces of feed, costing between 8 and 16 cents, are required to produce a dozen eggs. With egg prices of $.90 to $1.40 per dozen, there is a wide margin to cover other fixed and variable costs and still make a profit.

Quail are raised in paired cages, group cages, wire floor pens or litter floor pens. Wire floor units are often used because separation of the birds from their manure reduces disease problems.

Squab
A squab is a young pigeon. Squab can be a profitable source of income with good management and a good market. Investigate the market before seriously considering production.

Squab raising is compatible with other farming ventures. It provides regular income and the unique taste and texture of the meat makes it very marketable with a good return per acre.

Simple housing consists of pens with feeders, waterers, roosts, and nests. These pens are often open on three sides and screened with poultry netting. The nests are built on the solid back wall. Normally four to five square feet is allowed per pair, with up to 20 pair per pen.

Squab are processed at 28 to 30 days of age, when the pinfeathers under the wing are in their sheaths and not quite open. A pair of squabbing pigeons incubate and feed the squabs.

With good management a squabbing pair will produce 10 to 13 squab per year, valued at $30 to $40. Variable costs are about $25 per pair and fixed costs $8 per pair. This results in an estimated net income of up to $7 per pair. A unit of 500 squabbing pairs would require about 30 hours of labor per week, and, like all poultry raising, this is a 365-day per year job.

Game Bird Economics
Some farmers raise game birds, principally pheasants, partridges and quail, to sell to shooting clubs, wildlife managers, and landowners to increase resident

Enterprise Budget Game Birds (20 Weeks)

	Pheasant			Chukar Partridge			Bobwhite Quail		
		Total ($)	Per Bird ($)		Total ($)	Per Bird ($)		Total ($)	Per Bird ($)
Chicks (21,052)	$1.92 ea.	40,419		$1.84 ea.	38,736		$1.59 ea.	33,473	
Feed @ 10¢ pound	14.1 lb.	29,683	1.48	7.3 lb.	15,368	.77	4.5 lb.	9,473	.47
Labor (750 hours @ $11.50 per hour)		8,687	.43		8,687	.43		8,687	.43
Medication		755	.04		760	.04		778	.04
Insurance and taxes		1,532	.08		1,522	.08		1,555	.08
Depreciation and repairs		7,600	.40		7,600	.40		7,776	.40
Interest (10%)		9,256	46		7,653	38		6,565	.33
Utilities		766	.04		760	.04		778	.04
Misc. supplies		1,149	.06		1,140	.06		1166	.06
Advertising		766	.04		760	.04		778	.04
Management @.06/bird)		1,200	.06		1,200	.06		1,200	.06
Total		**$101,813**	**$5.09**		**$84,186**	**$4.21**		**$72,229**	**$3.61**

Additional Resources on Ostriches and Emus

American Ostrich Association,
4710 Bellaire, Suite 110,
Bellaire, TX 77401

American Emu Association
P.O. Box 9174,
Dallas, TX 75205

Bradley, Francine. Poultry Fact Sheets. Davis, CA: Department of Avian Sciences, University of California

Fitzhugh, Lee. Are Big Birds Good Business? Davis, CA: Department of Wildlife and Fisheries Biology, University of California.

Thornberry, Fred. Ostrich Production. College Station, TX: Texas A&M University

populations. Other game birds, including wild turkeys and guinea fowl, are sold mainly for meat.

There is an expanding market for game birds and the majority of the work can be done during the summer. But, the public often has a poor perception of pen-raised game birds and their management is very time consuming.

The cost study (p. 99) provides information on operating expenses for three popular farm-raised game birds—pheasant, chukar partridges, and bobwhite quail. The study is based on a 20,000-bird farm with a mortality rate of 15 percent (a very conservative rate). Pheasants are normally kept in range pens and chukar partridges and bobwhite quail are kept in colony cages, at a female to male ratio of three to one. Cost figures do not include profit margins for the sale of eggs, chicks, or marketable birds.

Ostriches and Emus

Ostriches are raised for their plumes, hides, and meat; however, currently, most stock is sold to other breeders. The hides are used to make boots, purses and accessories. The eggs and plumes are sold for decorating. The meat can be sold to upscale restaurants.

Producing ostriches is a very specialized business. If you consider such an operation, study production and care, and, as in other farm ventures, find buyers before you start production. Perhaps there is no market or it is too far away to make the business profitable. Perhaps the market is "soft" and the purchasers are dealing with a fad that won't last. If you think you can sell ostriches or their products, decide whether you have the expertise to handle the big birds. Do you have someone who will process the birds and hides for you? Is there a qualified vet in the neighborhood?

Ostriches are expensive. An adult pair may cost as much as $30-50,000. A four- to six-month old pair could cost nearly $6,000. Such an investment could mean borrowing or finding investors.

Emu raising is a growing enterprise. The birds are docile and easier to raise than ostriches. They usually weigh under 150 pounds as adults. The birds are as expensive as ostriches. They are sold for their meat, oil (good for cosmetics and therapeutics), and leather. Promoters describe their meat as tasting like beef but with fewer calories and cholesterol than chicken or fish. Contact the American Emu Association (see box) for more information.

Pigs

Everything you need to know about raising a small number of pigs is in the

Handbook for the Small-Scale Pork Producer. Subjects covered include marketing, site selection, housing requirements, health, and breed associations.

Currently, there is little market for once popular pot-bellied pigs.

Rabbits

Because they take little space, minimum investment, and are not labor-intensive, rabbits are often raised on small farms. Raising rabbits as a sole source of income is probably not feasible because the market is not big enough, but they are often part of mixed enterprise operations. They are used for their white, fine-grained meat for home use or for sale, pelts, feet (as charms), and manure. Live animals are sold to breeders, exhibitors and to laboratories.

Rabbits need clean, dry housing, protection from wind, and protection from weather extremes. They must be fed daily, have water available at all times, and have their cages cleaned regularly.

Refer to the UC publications for detailed information on breeds, selection, feeding, health, diseases, housing, reproduction, as well as butchering, and marketing of meat and pelts.

Goats

The market for goat products—milk, meat, fiber, processed products—as well as for breeding stock and show animals, is small and spread out.

Goat milk must sell at a premium since it competes with cow milk, so be sure of your market before committing your resources to the venture. Promotional programs, including information brochures and merchandise displays are vital to market development. The main customers are those who cannot tolerate cow products and some ethnic communities, but they are widely scattered, making distribution a problem. Farm sales may be more profitable. Consider processing the milk to increase its value. Goat cheese and goat milk ice cream are two products to consider. If you sell milk or meat products, you must follow strict health regulations.

Additional Resources and References on Pigs

Handbook for the Small-Scale Pork Producer. 1987. Pub. 21435. Oakland, CA: ANR Publications.

A Practical Guide to Swine Nutrition. 1977. Pub. 2342. Oakland, CA: ANR Publications. 8 p.

Swine Production. 1990. Pub. 21169. Oakland, CA: ANR Publications. 12 p.

Additional Resources and References on Rabbits

Common Rabbit Diseases. 1981. Pub. 2905. Oakland, CA: ANR Publications. 4 p.

The Rabbit Handbook. 1989. Pub. 21020. Oakland, CA: ANR Publications. 28 p.

Rabbit Reproduction. 1982. Pub. 2887. Oakland, CA: ANR Publications. 4 p.

Two-Unit, All-Wire Rabbit Hutch. 1983. Pub. 2737. Oakland, CA: ANR Publications. 4 p.

Additional References and Resources on Goats

American Cheese Society
34 Downing St.
New York, NY 10014; (212) 727-7939).
The American Cheese Society is a non-profit organization which encourages promotion of natural specialty cheeses. Membership includes a bimonthly newsletter and other literature.

Angora Goats. 1993. Washington, DC: USDA, CSRS. 2 p.

"Tri-County Dairy Goat Newsletter." Visalia, CA: Cooperative Extension.

Goats for Home Milk Production. 1976. Pub. 2899. Oakland, CA: ANR Publications. 4 p.

Nutrition and Feeding of Dairy Goats. Pub. WREP14. 1979. Oakland, CA: ANR Publications. 12 p.

Protecting Dairy Goats from Poisonous Plants. 1981. Pub. WREP46. Oakland, CA: ANR Publications. 4 p.

Your Dairy Goat. 1981. Pub. WREP47. Oakland, CA: ANR Publications. 28 p.

The market for mohair and cashmere is very small, undeveloped, and unorganized.

Often small farm families keep a few dairy goats because they are more convenient and economical for providing the family milk than dairy cows. Also, goats can be fun and friendly; can be trained and managed by youngsters for 4-H and FFA projects; can furnish the family milk, butter, cheese, and yogurt, and can help control brush.

If you are a beginner and want the goats mainly for milk, consider buying grade goats (parents of different or unknown breeds) rather than purebreds, because they cost less and can produce as well as registered goats. There is not much difference in milk production among different breeds. Nubians often give less milk than other breeds, but their milk is usually higher in butterfat. When selecting a dairy goat doe, check milk production, health and appearance, and herd health.

If you are interested in breeding and raising purebreds for sale and show, you should buy registered animals. The five main breeds of dairy goats in the US are: French Alpine, American La Mancha, Nubian, Saanen, and Toggenburg. Different breeds have different characteristics, whether it be hair color, milk composition, or meat quality.

Generally, kids are cheaper. New goat owners often start with kids so they can get acquainted with them and their habits as they grow. This gives the goat owner more time to learn about goats and how to manage them. It is usually better to have two animals because they provide company for each other. A single animal will be noisy and lonely.

Goats need clean, dry shelter free from drafts and outdoor space for exercising. They can stay in a shed or in stalls. Each goat needs 15 square feet of bedded area in a shed. The ceiling should be at least 10 feet tall so you can clean the stall with a tractor. The feeding area should be on one side of the bedded area since the animals will spend a lot of time there, and it may require daily cleaning.

During the summer a goat needs pasture and one to one-and-one-half pounds of grain per day. During the winter a dairy goat consumes two to four pounds of hay and one to two pounds of grain daily. Goats also eat beet tops, beans, sweet corn stover and other garden wastes. Dairy goats may have different feed requirements than non-milking goats, but both need a mixture of grain, hay and pasture.

A good dairy goat produces two quarts of milk daily for eight to 10 months of the year and can be fed for much less than a dairy cow can. Each doe produces one or two kids per year. Doe kids can be raised for replacements and buck kids can be raised for meat.

As in any animal enterprise, you must have the time to give the goats regular care. If you raise dairy goats, you will need to milk twice a day, every 12 hours. Cleanliness is extremely important. The milk needs to be cooled and refrigerated. If it is retailed it may need to be pasteurized. You have to care for your goats and their milk every day throughout the year.

You'll have to deal with lice, worms, parasites, and diseases, including tetanus.

Sheep

Sheep are raised for their wool and their meat. The best sheep for wool are not always the best for meat.

Sheep are marketed through word-of-mouth, newspaper ads, to 4-H and FFA youth, at auctions, and to other producers. The wool is sold to home spinners as well as commercial and local wool buyers.

Raising sheep and lambs is a full-time, year-around job.

Horses

Many small farmers keep one or more horses. They are usually pets used for the pleasure of the family. To make money, some innovative farmers board or breed pet horses, exotic horses, work horses, miniature horses, race horses, etc. Written information is abundant. Ask your neighbors and farm advisor for names of people selling horses, horse organizations, and other recommendations.

> ### Additional References and Resources on Sheep
>
> A Handbook for Raising Small Numbers of Sheep. 1985. Pub. 21389. Oakland, CA: ANR Publications. 56 p.
>
> Kruesi, William K. 1985. Sheep Raiser's Manual. Charlotte, VT: Williamson Publishing. 288 p.

Cows

Beef or dairy cows are often kept on small farms for family use or for production of income. They are sometimes kept on land that is unsuitable for other uses. Of course, large animals need room to graze and adequate housing, a large amount of food, and an owner who is physically able to handle the beasts. Caring for a cow is a 365-day-a-year job.

Some farmers increase their profits by producing organic beef (which needs organic feed and is not given hormones or antibiotics) or breeding exotic cattle. These animals bring a higher price. They are also more expensive to raise because the organic cattle need organic feed, which costs more, and the exotics cost much more to purchase as calves. Consumers are willing to pay more for organic beef—up to a point. To make money selling a specific breed depends on availability of the breed and your marketing ability.

Additional References and Resources on Cows

Beef Production in California. 1980. Pub. 21184. Oakland, CA: ANR Publications. 24 p.
The basics of cattle raising for prospective or new owners, small herd owners, and 4-H program participants.

Small-Scale Beef Raising: Reward or Risk. 1981. Pub. 2211. Oakland, CA: ANR Publications. 8 p.
For newcomers in the industry; problems involved in small operations.

Beekeeping

There are various ways to make money from keeping honey bees. Most beekeepers rely on honey and beeswax production and commercial crop pollination. Costs to start a beekeeping business are not particularly high compared to many small businesses, and a well planned and managed operation can be profitable.

Beekeepers own, rent, or find free apiary locations where their bees can forage for food while not becoming a nuisance to humans or livestock. Beekeepers must manage their colonies to the benefit of the bees and in compliance with existing state, county, or municipal beekeeping ordinances. In some areas hives can be left on one spot permanently. In most areas, suitable

Lamb Marketing

By Jeanne McCormack, farmer, Rio Vista area

My husband, Al Medvitz, and I began farming six years ago in Rio Vista, in the Sacramento Delta. We work for my father and help him with his 800 ewes, and we rent the adjacent ranch and raise 750 of our own ewes. We have developed a year-around lamb marketing business and sell about 1,500 lambs a year, or 30 to 35 a week. The approach we took to starting our marketing business applies to the small producer equally well as to a medium-sized commercial producer like us.

The first rule is to start small. I began with 40 lambs and my mother's Christmas card list and grew from there. We sold the first bunch of 40 lambs in a month—to individual households and to two up-scale restaurants in the San Francisco Bay area. I delivered cut, wrapped, frozen, and boxed lambs to householders and unfrozen carcasses to the restaurants. We gradually increased our direct marketing, beginning with five a week to a restaurant and ultimately selling through a restaurant distributor.

The second rule is to define clearly both your product and your market. Our product is naturally-grown, hormone- and antibiotic-free lamb, perfectly finished and with a delicate flavor. Our market is well-to-do consumers who either dine at expensive restaurants and expect to be served high quality food or are interested in gourmet cooking at home, want the best possible food stuffs, and are willing to pay a premium price for them.

The third rule is to develop a product of uniform quality, which is very difficult when you are raising animals. You don't know whether they taste good until you eat them, so quality control can be terribly expensive. We have a complicated breeding program and also must buy feeder lambs from other producers in order to meet the demand for our lambs. The feeder lambs stay on our ranch for several months and consume the same ration as our own lambs. Our buyer samples our lamb every week and discusses with us the flavor, texture, and toughness. We also regularly talk to restaurant chefs serving our lamb in order better to learn how they define "excellence" and to understand what in our system has produced either superior or inferior meat. The issue of quality control is particularly important with lamb, since many people do not like its flavor. We try to avoid a lamby or "woolly" flavor.

The fourth rule is to work out the logistics of butchering beforehand. Lamb sold to restaurants must be slaughtered at a federally-inspected packing plant, and there are not many of these. Lamb sold to individual householders must also be inspected unless it is sold live to the buyer, who then takes care of the butchering him or herself. You should do the research on how to get your lambs slaughtered and the meat prepared well before the lambs are ready to eat. You can learn what you need to know by calling your state meat inspection offices as well as interviewing people who operate local packing houses and ranch-slaughter businesses.

With adequate research and a proper definition of your market, you can find a profitable way to raise and sell your lambs.

permanent locations do not exist and the apiaries are relocated six or more times a year.

Bees are essential for pollination of more than 50 California crops, so farmers without their own bees rent from a beekeeper. Colony rental prices vary with supply and demand. Vine seed pollination in California in the summer is extremely inexpensive to the growers because beekeepers are searching for safe havens where their bees might get a little food. Rental rates are two to three times higher for almonds, which need to be pollinated when bee supplies barely meet demand.

Part-time beekeepers operate 50 to 450 colonies. Full-time beekeepers operate 1,000 and 4,000 hives. One knowledgeable beekeeper is needed for every 500 to 1,000 colonies.

Beekeepers need a place for storing and repairing equipment, mixing bee feed and antibiotic treatments, extracting and handling honey. Some people use a garage or toolshed, but a larger facility really is required. Many beekeepers rent buildings. Others build a "honey house." Before building a facility, visit other beekeepers to note features that make handling of equipment and honey efficient.

Beekeeping can sound deceptively simple. However, beekeeping is a form of animal husbandry involving providing feed when nectar and pollen are lacking, preventing infections by various microbes, dealing with two recently introduced parasitic mites, and reducing the influence of Africanized honey bees. Before trying to keep bees commercially on your own, you should gain experience working with a commercial beekeeper for awhile.

Aquaculture

Aquaculture, one of the fastest growing segments of American agriculture, includes raising fish to sell to markets and restaurants (live or processed) as well as on-farm sales, including recreational

Investment Needed for 1,000 Colony Operation

Hive Equipment

1,000 bottom boards @ $8 each	$ 8,000
1,000 covers @ $8 each	8,000
2,000 deep boxes @ $12 each	24,000
20,000 deep frames @ $0.35-0.65	10,000
20,000 deep foundation @ $0.06	1,200
1,000 medium depth boxes @ $8 each	8,000
10,000 medium depth frames @ $0.40 each	4,000
10,000 medium depth foundation @ $0.40	4,000
100,000 frame eyelets @ $2.00 per 1,000	200
2,000 queen excluders (optional) $9.00 each	18,000
6,000 metal rabbets @ $0.08 each	480
50 fume boards @ $9.00	450
1 bee blower (optional) @ $325 each	250
75 gallons paint @ $16-21 per gallon	1,500
1 staple gun and compressor	500
Bees 1,000 packages @ $25.00	25,000

Honey Handling Equipment

Automatic uncapper	1,700-3,000
Frame conveyor	600
Conveyor drip pan	250
Cappings melter	1,000-2,000
Extractor	1,900-7,800
Settling tanks (each)	170-250
Spin float (replaces melter)	3,300
Honey sump	325-800
Honey pump	170-190
Flash heater (optional)	1,000
Barrels (each)	new: 16; used: 8
Barrel truck	160-250
Hand truck	125-525
Glass jars (if not selling bulk honey)	17,300
Bottling equipment (if not selling bulk honey)	940

Vehicles

Flat bed trucks (each)	16-18,000
Bee booms (each) (mounted)	2,500
Forklifts (each)	new: 16-18,000 used: 8-10,000
Pickups	14,000
Warehouse	6,000
Land @ $3,000/acre	20,000
Rent (house and shop/year)	15,000-17,000

Labor

Self	30,000
Help, full time, each	20,000
Help, part time, each	1,630

Overhead

Utilities (year)	2,400
Insurance	varies
Workman's compensation, health insurance	13,000

Additional Resources and References on Beekeeping

Beekeeping in California. 1987. Pub. 21422. Oakland, CA: ANR Publications. 72 p.

Graham, Joe (editor). 1992. The Hive and the Honey Bee. Dudant and Sons. 1,324 p.

Sammataro, Diana and Alphonse Avitabile. 1986. Beekeeper's Handbook (2nd ed.). New York, NY: Collier Books. 148 p.

fishing lakes. Fingerlings are sold to government and private groups for lake and stream replenishing. A commercial aquaculture operation is a full-time venture.

To participate, you need good water quality and water and land that match the specific growth, health, and reproduction requirements of the species you choose to raise. For instance, trout do best in water temperatures of about 50-65°F, while catfish thrive at 84°F. The temperature of the water can be manipulated, but the expense may not be worthwhile. Have your water analyzed to determine whether it is or could be made suitable for aquaculture (see chapter 2 for water analysis information). Also evaluate your water volume, velocity of flow, and exchange rate.

Gather information and ask questions about marketing, financing, government regulations and permits. Study availability and costs of labor, water, energy, and feed. Your costs will also depend on your distance from suppliers and markets. The worksheet at the end of the book will help you analyze your situation. Some aquaculturists start with small investments and, if successful, make larger commitments.

You don't have to own a large tract of land to consider building or modifying a pond for aquaculture. The Soil Conservation Service will help you plan your pond, including site examination, soil analysis, surveying, blueprint preparation, writing up pond and dam specifications, and listing materials needed. (They are listed under USDA in the government section of your phone book.)

The amount you can charge for your fish will fluctuate constantly. The price will depend on non-domesticated harvest, food safety concerns, foreign aquaculture production, competition from red meat and poultry, resource constraints, and your marketing strategies. Maintaining quality control is vital.

Five of the most popular cultured fish in California are catfish, trout, striped bass, sturgeon, and oysters. Catfish is usually sold live in fish markets. (The catfish fillets you see at your supermarket are probably from Mississippi. There are no catfish processing facilities in California.) Most trout are used to stock lakes. Sturgeon and oysters are usually sold processed.

Fingerling catfish

Aquaculture References and Resources

California Aquaculture Association
P.O. Box 1004
Niland, CA 92257.

Most of these publications are available through your local Cooperative Extension office, or they will help you obtain them.

Conte, Fred. Aquaculture Is Agriculture, State Laws which Stipulate that Aquaculture is a Form of Agriculture. Pub. ASAQ-B3-7/90. Davis, CA: Department of Animal Science, University of California.

Conte, Fred. Aquaculture Sources, Producer Associations, Aquaculture Agencies, Institutions and Services. Pub. ASAQ-B7-10/90. Davis, CA: Department of Animal Science, University of California.

Conte, Fred. Channel Catfish, Commercial. Pub. ASAQ-A1-8/90. Davis, CA: Department of Animal Science, University of California.

Conte, Fred. Chemical Analysis Data Sheet for Evaluation of Water for Aquaculture Potential. Pub. ASAQ-C1. Davis, CA: Department of Animal Science, University of California.

Conte, Fred. Evaluating Resources. Pub. ASAQ-C8-10/90-4/92. Davis, CA: Department of Animal Science, University of California.

Conte, Fred. 1993. Evaluation of a Freshwater Site for Aquaculture Potential. WRAC Publication 92-101. 35 p.

Conte, Fred. Periodicals, Books, Marketing Associations. Pub. ASAQ-B8-10/90. Davis, CA: Department of Animal Science, University of California.

Conte, Fred. Permit Guide, California Lead Agency for Aquaculture and Assistance with the Permit Process. Pub. ASAQ-B1-7/90. Davis, CA: Department of Animal Science, University of California.

Conte, Fred. Suggested Water Quality Criteria for Aquaculture Hatcheries or Production Facilities. Pub. ASAQ-C9. Davis, CA: Department of Animal Science, University of California.

Trout Ponds for Recreation. Farmers' Bulletin Pub. 2249. Washington, DC: USDA. 13 p. Site selection and preparation, stocking, feeding, harvesting, problems.

Warmwater Fish Pond Management in California. Washington, DC: USDA Soil Conservation Service.

Additional Resources and References

Exotic Livestock, a Small-Scale Agriculture Alternative. Washington, DC: CSRS, USDA. 2 p.

Wessel, Kelso L., Debra Britton, Rebecca Boerger. 1991. Innovative Rural Enterprises, The Farm Income Enhancement Program. Columbus, OH: Ohio State University.

IX
POSTHARVEST HANDLING OF PERISHABLE CROPS

Effective postharvest management, rather than any given technology, is the key to maintaining quality (appearance, texture, flavor, nutritive value and safety) and reducing losses of fruits and vegetables between harvest and consumption. Small-scale handlers cannot afford costly machinery and high-tech postharvest treatments. But, simple, low cost techniques—minimizing rough handling, sorting to remove damaged and diseased produce, and effective temperature and humidity management—go a long way toward maintaining quality and reducing losses.

Harvesting and Preparation for Market

Damage during harvest predisposes produce to decay, water loss, and increased respiratory and ethylene production rates, which lead to quick deterioration. Harvesting by machine causes more damage than harvesting by hand, although some root crops can be severely damaged by careless hand digging. Field containers should have smooth inside surfaces, be free of rough edges, and be clean. Vented, stackable plastic crates, while initially expensive, are durable, reusable, and easily cleaned.

Field workers should harvest crops to minimize damage and waste, and should be trained to recognize when the fruit or vegetable is ready to pick. They should snap, cut, or pull the fruit or vegetable from the plant in the least damaging manner. Knife

Chapter Authors: Adel A. Kader, Professor, Department of Pomology, UC Davis; Lisa Kitinoja, Consultant, Extension Systems International, Woodland, CA tips should be rounded to minimize inadvertent gouges on perennial plants. Knives and clippers should be sharp. Pickers should empty their bags with care, never throwing produce into containers. If they pick directly into bins, bruising can be minimized by using a canvas de-accelerating chute. Field containers should be clean and smooth.

Protect produce from exposure to the sun after harvest. Place field bins in the shade or loosely cover them if they can't be removed from the field immediately. Consider night or early morning harvest to protect produce from heat and to reduce the energy needed for subsequent cooling.

Cooling—removing field heat directly after harvest and before further handling—lengthens postharvest storage life and helps maintain quality. Even produce repeatedly cooled and warmed deteriorates more slowly than produce that has not been cooled at all.

Rough handling increases bruising and other damage and limits the benefits of cooling. Secure field boxes during transport and, if stacked, do not overfill. Adjust transport speed to the quality and conditions of the roads, and keep truck and trailer suspensions in good repair. Reduced tire air pressure reduces the amount of motion transmitted to the produce.

Asian vegetables packed for shipment

Field packing (selecting, sorting, trimming, and packaging) greatly reduces the amount of handling the produce undergoes before marketing. Small, mobile field packing stations can be designed to be moved along with the pickers. Any practice that reduces the number of times the produce is handled will help reduce losses.

Packing

Packinghouse operations can include dumping, sorting, cleaning, waxing, grading and sizing, and packaging. Shade during packing is extremely important, whether from a plastic or canvas sheet hung from a pole or a permanent roofed structure. When deciding where to locate a packinghouse, consider access to the field, access to the market, space for vehicles to enter and leave, and ease of access for laborers.

Dumping from field containers to assembly line packing can be either dry or water assisted, depending on the kind of produce being handled. Water assisted dumping can reduce bruising and is often used with apples and pears.

Sorting eliminates injured, decayed, or otherwise defective produce (culls). This saves energy because the culls will not be cooled. Removing decaying produce also limits the spread of infection, especially if postharvest pesticides are not used.

Produce may be cleaned by washing in chlorinated water or dry brushing. For some commodities, such as kiwifruits and avocados, dry brushing is sufficient to clean the produce. Other commodities require washing. The choice of brushing or washing depends on the type of commodity and the type of possible contamination. Sanitation is essential to control the spread of disease and to limit spore buildup in wash water or in the packinghouse air. Chlorine in the wash water helps control pathogen buildup (one to two ounces of chlorine bleach per eight gallons). You can clean walls, floors, and packing equipment with quaternary ammonium compounds labeled as safe for food processing equipment.

Waxing is done after washing and air drying. Waxing of immature produce, such as cucumbers and summer squash, and mature produce, such as eggplant, peppers and tomatoes, replaces some of the natural waxes removed in cleaning and helps reduce water loss during handling and marketing. Allow the wax coating to dry thoroughly before further handling.

Grading and sizing separate the produce into processing and fresh market categories. The best produce is packaged and marketed rather than processed. Sizing produce may be worthwhile if certain size grades receive a higher market value than others. Mechanical sizers composed of a series of conveyors fitted with belts with various sized openings are available for most commodities. A simple method for mechanical sizing is to use a set of diverging bar rollers, where the smallest sized produce falls through the rollers to a sorting belt or bin earlier than larger sized produce. Most low-input packinghouses size manually. The packer selects the size desired and either packs the items into containers or places the produce gently into a bin for packing later. It helps the packer if examples of the smallest and largest acceptable produce are available.

Packing Materials

Heavy, waxed cartons or plastic containers are better for packaging produce than bags or open baskets because they protect the produce when it is stacked. Packaging also allows quick handling and minimizes impacts of rough handling. Packages need to be vented, yet sturdy enough so they will not collapse.

Waxed cartons and plastic containers, while more expensive, are reusable and stand up to the high relative humidity found in storage. They should not be filled either too loosely or too tightly. Shredded newspaper is an inexpensive and lightweight filler.

Packaging can be both an aid and a hindrance to storage life and quality. Packing materials protect the commodity by immobilizing and cushioning it, yet they may block ventilation holes, which has the undesirable effect of raising humidity in the package and preventing cooling. Packaging in plastic films can modify the atmosphere surrounding the produce by restricting air movement and allowing the buildup of carbon dioxide.

Decay and Insect Control

Careful harvesting, sorting out damaged or decaying produce, and carefully preparing produce for market helps prevent insect and disease damage and decay. Yet, even if the greatest care is taken, sometimes produce must be treated.

While high humidity in the storage environment is important for maintaining high quality produce, water on the surface of commodities can invite pathogens. When cold commodities are removed from storage to higher temperatures, moisture from surrounding air can condense on the product. Using a fan can help reduce the chances of infection.

Jicama

Certain fungi in their germination phase are susceptible to cold, and infections can be reduced by treating produce with a few days of storage at the coldest temperature the commodity can withstand without incurring damage.

Cold treatments can control some insects, including fruit flies. Since the treatment requires 10 days at 0° C or below, or 14 days at 1.7° C or below, it is most suitable for commodities adaptable to those conditions, such as apples, pears, grapes, kiwifruit, and persimmons.

Short hot water dips or forced-air heating can be effective against certain fungi on plums, peaches, cantaloupe, stone fruits, sweet potatoes, and tomatoes.

Hot water dips or heated air can be used for direct control of postharvest insects. Fruit should not be handled immediately after heat treatment. Whenever heat is used with fresh produce, cool water showers or forced cold air should be used to return the fruit to a low temperature as soon as possible.

Control of insects in nuts and dried fruits and vegetables can be achieved by freezing, cold storage (less than 5° C), heat treatments, or the exclusion of oxygen (0.5 percent or lower) using nitrogen. Packaging in insect-proof containers prevents subsequent insect infestation.

Washing produce with chlorinated water can prevent decay caused by bacteria, mold and yeasts. Calcium hypochlorite (powder) and sodium hypochlorite (liquid) are inexpensive and widely available. The treatment is less effective if organic matter is allowed to build up in the wash water.

Some chemicals used to control molds, fungi and insects on fruit and vegetable crops are sulfur, sodium or potassium bisulfite, alum and lime powder and pesticides. The concentrations, method, and length of treatment depend on the crop and the problem.

Temperature and Relative Humidity Control

From harvest to consumption, temperature control is the most important factor in maintaining quality. Fruits, vegetables and cut flowers are living, respiring tissues separated from their parent plant. Keeping them at their lowest possible temperatures (0°C for temperate crops or 10-12°C for chilling sensitive crops) will increase their storage life by lowering respiration rates, decreasing sensitivity to ethylene, and reducing water losses. It is important to avoid chilling injury, since this can lead to failure to ripen (bananas and tomatoes), development of pits or sunken areas (oranges, melons and cucumbers), brown discoloration (avocados, cherimoyas), increased susceptibility to decay (cucumbers and beans), and development of off-flavors (tomatoes).

Refrigerators are the most reliable method of chilling. Other systems include night air ventilation, radiant cooling, evaporative cooling, the use of ice, and underground or high altitude storage.

Simple practices, such as shade over harvested produce, buildings, and vehicles, enhance storage efficiency. Trees are a fine source of shade and can reduce the temperature around packinghouse and storage areas. Light colored buildings reflect light and reduce heat. Sometimes spending money saves money, such as purchasing high pressure sodium lights, which produce less heat and use less energy than incandescent bulbs.

Also, consider the relative humidity of the storage environment. Loss of water from produce is associated with loss of quality, because visual changes—wilting, shriveling, textural changes—occur. However, water loss may not always be undesirable; for example, if produce is destined for dehydration or canning.

For fresh market produce, any method of increasing the relative humidity in the storage environment slows water loss. One method is to reduce temperature; another is to add moisture to the air around the commodity using mists, sprays, or wetting the storeroom floor. Or, use vapor barriers such as waxes, polyethylene liners in boxes, or coated boxes. Packaging materials can interfere with cooling, so vented liners (about five percent of the total area of the liner) are recommended.

Room Cooling

Room cooling is a low cost but slow method for cooling produce. The greater the refrigerator coil area, the less moisture will be lost as the product cools.

Leave adequate space between stacks of boxes inside the refrigerated room so produce can cool quickly. Stacks should be only one pallet width deep. Cooling occurs as air circulates through the room, over surfaces and through open spaces.

A low cost cold room can be constructed with concrete floors and polyurethane foam as an insulator. A square shape reduces the surface area per unit volume of storage space, thereby reducing construction and refrigeration costs. Rectangular buildings have more wall area per square foot of storage space, and are therefore more expensive to cool. Caulk joints and put a rubber seal on the door.

Forced-Air Cooling

Forced-air cooling pulls air through storage containers, greatly speeding the cooling of produce. A variety of forced-air coolers are available for use inside cold rooms (fixed or portable models). Some forced-air coolers move cold, moist air over the commodities.

Hydrocooling

Hydrocooling provides fast, uniform cooling for some commodities, and helps avoid water loss. The commodity as well as its packaging materials must be tolerant of wetting, chlorine (used to sanitize the hydrocooling water) and water beating damage. Crops typically hydrocooled include root vegetables, asparagus, celery, rhubarb and sweet corn.

The simplest version of a hydrocooler showers a batch of produce with icy water. Water flows of 15 to 25 gallons per minute per square foot of surface area are generally desirable. The water should contain 100 ppm chlorine for best results. Typical cooling times are 10 minutes to one hour depending on the size of the product. A batch-type hydrocooler can be constructed to hold entire pallet loads of produce. Conveyors can be added to help control the time produce stays in contact with the cold water.

Use of Ice

The use of ice to cool produce provides a high relative humidity environment around the product. Ice can be a bunker source of cold (used by passing air through a bank of ice and then through the commodity) or be used as top ice (laid directly in contact with the product). Ice can cool a commodity only if it melts, so good ventilation is necessary for effective cooling.

Top ice can be used only with water tolerant, non-chilling sensitive products (for example: artichokes, peas, carrots, sweet corn, cantaloupes, lettuce, spinach, broccoli, green onions), and with water tolerant packages (waxed cardboard, plastic or wood). Crushed or flaked ice can be applied directly or as a slurry in water.

Storage

If produce is to be stored, begin with a high quality product. There must be no damaged or diseased pieces and containers must be well ventilated and strong enough to withstand stacking. Proper storage practices include temperature control, relative humidity control, air circulation, maintaining space between containers for ventilation, and avoiding an incompatible product mix. Some produce (potatoes, sweet potatoes) should be cured before long-term storage. Curing is accomplished by holding the produce at high temperature and high relative humidity for several days while harvesting wounds heal and a new protective layer of cells forms.

Commodities stored together must tolerate the same temperature, relative humidity and level of ethylene. High ethylene producers (such as ripe bananas, apples, cantaloupe) can stimulate physiological changes in ethylene sensitive commodities (such as lettuce, cucumbers, carrots, potatoes, sweet potatoes) leading to color, flavor and texture changes.

It is easier to manage temperature during storage in a square rather than a rectangular building. Shaded buildings, white or silver paint, and roof sprinklers also help keep the temperature down. Ventilation in storage structures is improved if air inlets are located at the bottom of the facility, and air outlets are at the top. Use of a simple exhaust fan or an evaporative cooling unit is often adequate for chilling sensitive crops. Buildings set below ground level or partially covered with soil stay cooler than storehouses build above ground.

Onions and garlic keep better at a lower relative humidity. Allowing the external layers of tissue to dry prior to handling and storage protects them from further water loss.

Transportation

Temperature management is critical during long distance transportation. Produce must be carefully stacked, braced and well secured to minimize damage. Stack loads on pallets and away from the walls of the vehicle, leaving several open channels between stacks to allow proper air circulation to carry heat away from the produce. Vehicles should be well insulated to maintain cool environments for pre-cooled commodities and well ventilated to allow air movement.

Mixed loads are a serious concern when optimum temperatures are not compatible (for example, when transporting chilling sensitive fruits with other commodities) or when ethylene producing commodities and ethylene sensitive commodities are transported side by side.

Handling at Destination

Before produce is sold to the consumer, the handler may need to sort for quality, or at least to discard any damaged or decayed produce. If ripeness or maturity is non-uniform, sorting at destination can provide the seller with a higher price for the better quality produce.

When displaying horticultural crops, single or double layers of produce are most likely to protect the commodities from compression damage and over handling by consumers.

Outdoor marketplaces suffer from a lack of temperature control and high air circulation, which can lead to quick desiccation of crops (shriveling and wilting). These markets can often benefit from the increased use of shading and protection from prevailing winds. Misting commodities that can tolerate surface water (lettuce, broccoli, green onions) with cool, clean water can help maintain a high relative humidity around the product.

Finally, the handler at destination can help reduce losses in the future by maintaining good records of the sources of losses suffered at the wholesale or retail level. Identifying whether losses were due to mechanical damage, decay/disease, immaturity or over-ripeness allows the handler to provide better quality feedback to produce suppliers.

Additional Resources and References

Kitinoja, Lisa and Adel A. Kader. 1994. Small-Scale Postharvest Handling Practices: A Manual for Horticultural Crops. Postharvest Horticulture Series No. 8. Davis, CA: Department of Pomology, University of California. 199 p.

Hardenburg, R.E., et al. 1986. The Commercial Storage of Fruits, Vegetables, and Florist and Nursery Stocks. Washington, DC: USDA. 130 p.

Kader, Adel A. (ed). 1992. Postharvest Technology of Horticultural Crops (2nd edition). Pub. 3311. Oakland, CA: ANR Publications. 296 p.

Kader, Adel A. et. al. 1994. Postharvest Resources Directory. Postharvest Horticulture Series No. 5. Davis, CA: Department of Pomology, University of California. Sources of packaging materials and other postharvest supplies.

X
ALTERNATIVE AGRICULTURE

Alternative agriculture is any system of food or fiber production that systematically pursues the following goals:

- More thorough incorporation of natural processes such as nutrient cycles, nitrogen fixation, and pest-predator relationships into the agricultural production process;

- Reduction in the use of off-farm inputs with the greatest potential to harm the environment or the health of farmers, farm workers, and consumers;

- Greater productive use of the biological and genetic potential of plant and animal species;

- Improvement of the match between cropping patterns and the productive potential and physical limitations of agricultural lands to ensure long-term sustainability of current production levels; and

- Profitable and efficient production with emphasis on improved farm management and conservation of soil, water, energy, and biological resources.

  From: National Research Council, 1989, Alternative Agriculture, p. 27.

Chapter authors: William Burrows, Instructor, Shasta Community College, Redding; Shirley Humphrey, Staff research Associate, Small Farm Center, UC Davis; Karen Klonsky, Extension Economist, Agricultural Economics, UC Davis; Pete Livingston, Staff Research Associate, UC Davis; Eric Mussen, Extension Apiculturist, Entomology Extension, UC Davis; Claudia Myers, Associate Director, Small Farm Center, UC Davis; Richard Smith, Farm Advisor, Monterey, San Benito, Santa Cruz Counties; Laura Tourte, Postgraduate Researcher, UC Davis; Suzanne Vaupel, Attorney, Agricultural Economist, Vaupel Associates, Sacramento

Alternative agriculture includes farming systems ranging from those which use no purchased synthetic chemical inputs to farming systems that allow prudent use of pesticides or antibiotics to control specific pests or diseases. Alternative farming systems include sustainable, organic, biological, biodynamic, low-input, regenerative, holistic, etc.

Farmers are increasingly embracing alternative systems as they and their consumers become concerned about health problems and the short- and long-range costs of water and soil pollution attributed to excess chemical use, and soil depletion and erosion.

> The US Environmental Protection Agency has identified agriculture as the largest nonpoint source of surface water pollution. Pesticides and nitrate from fertilizers are detected in the groundwater in many agricultural regions. Soil erosion remains a concern in many states. Pest resistance to pesticides continues to grow, and the problem of pesticide residues in food has yet to be resolved. Purchased inputs have become a significant part of total operating costs. Other nations have closed the productivity gap and are more competitive in international markets.
>
> Because of these concerns, many farmers have begun to adopt alternative practices with the goals of reducing input costs, preserving the resource base, and protecting human health.
>
> The hallmark of an alternative farming approach is not the conventional practices it rejects but the innovative practices it includes. In contrast to conventional farming, however, alternative systems more deliberately integrate and take advantage of naturally occurring beneficial interactions. Alternative systems emphasize management; biological relationships, such as those between the pest and predator; and natural processes, such as nitrogen fixation instead of chemically intensive methods.
>
> From: National Research Council, 1989, Alternative Agriculture, p. 3.

Sustainable Agriculture

Sustainable agriculture integrates three main goals—environmental health, economic profitability, and social and economic equity. Farming methods include:

- Using natural processes, such as nitrogen fixation with cover crops; biological pest and disease control with IPM, natural enemies, and tolerance for some weeds and insects; and mixing low-density animal production with plant production.

- Reducing (but not necessarily eliminating) purchased inputs for synthetic chemicals by substituting IPM and increasing tolerance for pests.

- Understanding a plant's or animal's genetic potential or limits, and how it interacts with its environment.

- Achieving a profit for the farm operator and adequate wages for farm workers.

- Conserving, restoring, and protecting natural resources through such practices as IPM, crop rotation, conservation tillage, improving soil fertility with cover crops, eliminating ground and water pollution, and, most of all, thoughtful management.

Sustainable farming may be conducted on a large or small scale, on a single crop or mixed farm, using organic or conventional inputs and practices. Sustainable practices are particularly attractive to small-scale farm operators who may have more time and know-how than money.

Sustainable agricultural production is not solely concerned with high yield. In fact, yield may be lower, but because inputs are less expensive, profit may not be sacrificed.

Land stewardship is an important aspect of sustainable agriculture.

Organic Farming

J.R. Organics greenhouse

Sir Albert Howard, a British agronomist, popularized organic farming methods in the 1940s. Growing plants and husbanding animals without using synthetic chemicals was vital in the economically depressed area of India where he worked. The farmers had to recycle natural nutrients from waste products because they couldn't afford any purchased inputs. Sir Albert was disturbed by newly emerging "scientific" farming because the rhythms of nature that had built the soil were being ignored. He taught farmers to return wastes to the soil, combat insects without poison, and avoid synthetic fertilizers and their toxic residues. J.I. Rodale continued Sir Albert's work through the Rodale Institute.

Regulations Governing Organically Grown Produce
The term "organic" has legal implications as well as biological and philosophical. To call your product "organic" you must comply with these laws. If you don't, stiff penalties can result. The laws require extensive record keeping. There are strict prohibitions on what can be applied to the product and how the product is produced. California requires you to register with your County Agriculture

Escondido growers Joe and Carlos Rodriquez

Commissioner and pay a fee; federal law requires certification by an accredited organization. (Contact California Department of Food and Agriculture, Organic Program, for information.)

Approved and prohibited materials: Both California and federal law prohibit the use of any synthetic substance in the production of organic food products, with exceptions for specific materials. Each law lists the types of exceptions. Under California law, you may not use prohibited materials for 12 months prior to certification. That is: 12 months prior to appearance of flower buds for perennial crops; 12 months prior to planting or transplanting for annual or two-year crops; 12 months prior to seeding or inoculation of the medium for crops grown in a growing medium; or 12 months prior to germination of seed for crops grown without soil or growing medium. Federal law will require a three-year time period.

Regulatory Agencies: The California Department of Food and Agriculture (CDFA) administers the California organic program for raw agricultural commodities and eggs, and for raw and processed meat, fowl, and dairy products. The California Department of Health Services (DHS) administers the organic program for fish, seafood and processed foods other than meat, fowl, and dairy products.

The Transportation and Marketing Division of the United States Department of Agriculture (USDA) administers the national organic program.

Registration: In California all producers and handlers are required to register with the Agricultural Commissioner in the county of principal operation of the farm. Registration must be renewed annually. Registration is not required by federal law.

Certification of organic crops is voluntary under California law. Federal law requires every farm that produces food sold as organic to be certified. All certifying organizations must be accredited by USDA and CDFA. Certifying organizations must conduct an annual on-site inspection of each farm; require periodic residue testing; and recertify each farm annually.

Crossing Borders: Food grown outside California and sold in California as organic must meet all the same standards as required for food grown in California. Food grown in California and sold as organic outside the state must meet all the same requirements. Food grown outside the US and sold in the US as organic must be certified by a certifying organization that provides the same safeguards and guidelines required by US law.

Advantages and Disadvantages

One main advantage of using organic methods is the possibility of a premium price in the market place. Organic farmers can also save money over conventional farmers on purchased inputs like pesticides and herbicides. It also eliminates or reduces costs for pesticide training, equipment, posters, etc. Organic farmers can save money on irrigation due to improved soil tilth. Improved soil tilth and fertility result in less weeds and soil erosion. The disadvantages, particularly for the beginner, are that organic production takes longer, includes greater risk, and requires different knowledge or experience than conventional agriculture. Most farm management tasks are the same (pruning, irrigation, postharvest handling, etc.) whether you farm organically or conventionally. But to farm organically, you must understand the intricacies of organic pest control (particularly weeds) and fertility management.

Pest Management

Pest management may be the most challenging task for organic farmers. Losses due to pests can reduce yields or quality of the product. A more lenient market and price premiums for organic products are necessary to offset these problems.

Organic and conventional farmers challenge pests in many ways—crop rotations, diversification, spacing, timely planting, appropriate plant or seed and site selection, good water and soil management, mulching, mowing, or manual cultivation—but conventional farmers can rely mainly on chemical pesticides, while organic farmers attempt to control pests with more diversified methods. They sometimes use "botanical" insecticides and insecticidal soaps to control pests. Other methods of pest control include pheromone confusion and the use of beneficial insects. Farmers try to catch infestations early by inspecting crops frequently.

Weeds are largely managed by tillage with a sled or rolling cultivator. Hand weeding is often used in addition to mechanical cultivation. Flame weeders are also used.

Plastic or organic mulches block light and prevent weed seed germination. Vegetable crops are planted directly through mulches, which keeps weed populations low throughout the growing season. Because moisture retention generally increases with mulching, water use decreases. The use of some mulching materials may be restricted by registration or certification agencies. The best weed control in organic vegetable systems often results from the integration of a number of these techniques.

Marketing

Commodities that are produced organically can often be sold for more than conventionally grown products. However, the industry is extremely competitive. Profit is dictated by the amount of production, consumer demand, and the available organic market. Market saturation sometimes occurs. Growers are then forced to accept a lower price or market their products without the organic designation.

Although research indicates that organic vegetable production has expanded considerably in the last 20 years, there is some question about whether consumer demand and marketability have kept pace. A product's price, its appearance, and the consistency of supply influence a consumer's willingness to purchase organic commodities. The perceived benefits of organic agriculture, including food safety, improved nutritive values, and positive environmental impacts, also influence the decision of consumers to purchase organic products.

Organic Production Study

Recently UC researchers Karen Klonsky, Laura Tourte, and David Chaney, with contributions by Pete Livingston and Richard Smith, studied organic vegetable farms on the Central Coast. Farming is generally intensive—two or three crops are harvested from the same acreage each year. For each crop, cultural operations begin with land preparation. The soil is disked once or twice, then chiseled once or twice, and again disked once or twice, depending on the amount of residue to be turned under from the previous crop and the tilth of the soil. Land leveling is not commonly performed.

Soil amendments such as manure, composted manure, and gypsum are spread during land preparation to manage soil fertility and increase soil organic matter and nutrient levels for the next crop. Manure and composted manure provide nitrogen. Gypsum adds calcium and sulfur and improves soil tilth. Often soil amendments are added if warranted by appropriate sampling. Other soil fertility management techniques include crop diversification, rotation, and cover cropping.

After the ground has been worked and soil amendments added, beds are prepared. Weed seeds which have been brought to the surface are germinated with a preplant irrigation. A subsequent cultivation reduces or eliminates these weeds. Rolling and sled cultivators are commonly used.

Central Coast organic vegetable producers commonly plant on 40-inch beds. Vegetable crops are direct seeded, hand planted, or transplanted, depending on the crop, the time of year, the crop's ability to compete with weeds, and the targeted market. Organic growers must purchase and plant seed which has not been fungicide-treated; transplants must be grown in accordance with the California Organic Foods Act of 1990 and the Organic Foods Production Act of 1990. Organic growers should follow the guidelines of both Acts.

Most growers do not plant related crop species on the same acreage in any given year. This practice may be extended for longer time periods. Though initial land preparation is similar for most crops, different production practices—planting, irrigating, pest management, harvesting and packing—depend on the specific crops. For example, pest management materials may need to be applied to some vegetable crops while none may be required for other crops. Also, harvesting and packing methods may vary depending on available labor, equipment, or packing facilities.

Because of the importance of soil fertility and the need for high levels of soil organic matter, organic growers are increasingly planting some acreage to cover crops. Cover crops are beneficial for intensive organic vegetable production in a number of ways. Leguminous cover crops increase nitrogen in the soil. Water penetration and infiltration can be improved by root growth of the cover crop and by returning organic matter to soils. Increased organic matter can improve the soil's ability to retain moisture. Vegetables planted after a cover crop may not require any compost or manure applications. Weed suppression for subsequent crops is another benefit. Furthermore, cover crops increase plant diversity in the farming system and in the flowering stage can provide nectar to attract and sustain beneficial insects.

Yield and Return Ranges for Selected Central Coast Organic Vegetable Crops

	Yields Per Acre		Units	Returns Per Unit*	
	Low	High		Low	High
Cabbage (Green)	200	900	50 lb boxes	$4	$18
Cauliflower	600	1,000	12 count boxes	$4	$14
Cucumber	600	1,000	24 lb boxes	$4.50	$10
Garlic	300	500	30 lb boxes	$36	$68
Lettuce					
Leaf	350	850	24 count boxes	$5	$10
Romaine	300	750	24 count boxes	$5	$11
Onions					
Red	500	800	40 lb boxes	$5	$16
Yellow	600	1,000	50 lb sacks	$5	$16
Peas					
Snap	500	1,000	10 lb boxes	$5	$18
Snow	250	500	10 lb boxes	$5	$18
Peppers					
Green Bell	600	1,200	28 lb boxes	$4	$14
Red Bell	300	700	28 lb boxes	$5	$27
Sweet Corn	200	400	48 count boxes	$5	$12
Winter Squash					
Large Varieties	500	850	35 lb boxes	$5	$12
Small Varieties	200	850	20 lb boxes	$6	$13

*Figures represent returns to growers after cooling and/or marketing fees have been subtracted.
From: Klonsky, Karen, Laura Tourte, David Chaney, Pete Livingston, Richard Smith. 1993. Production Practices and Sample Costs for a Diversified Organic Vegetable Operation in the Central Coast. Davis, CA: University of California Cooperative Extension.

J. R. Organics
by Erin Chapman, Agricultural Consultant

Joe and Carlos Rodriquez own J. R. Organics, a 100-acre farm north of Escondido in San Diego County. On five acres they grow strawberries and vegetables (tomatoes, squash, cucumbers, cabbage, kale, beans, lettuce) and on 20 acres, flowers (baby's breath, marguerite daisies, sweet William, chrysanthemum, waxflower). The vegetables are grown organically, according to the requirements of California Certified Organic Farmers, which prohibit the use of synthetic chemicals and require a program to build soil fertility.

"At one time there wasn't a lot of chicken manure available, and the price of synthetic fertilizer was reasonable. So we used that instead. We noticed the soil was starting to get a lot of compaction, and there was no more organic matter left in the soil, and crop production started going down. So we went back to adding chicken manure after every crop," according to Joe.

Their pests include worms and aphids in the vegetables, and aphids, thrips, leafminers, and cucumber beetles in the flowers. Because the vegetables are certified organic, pest controls include biological insect control (releasing ladybugs and lacewings) and spraying a bacterium and insecticidal soap. The pests in the flowers are controlled with synthetic insecticides and fungicides. Joe says, "We noticed when we started using less pesticides and bringing in beneficial insects and not using any other category I materials, we had better control of pests . Before, we would use category I chemicals and lose control over insects. Using chemicals that aren't as strong is better because the insects don't build up a resistance. By alternating chemicals you get a better kill."

To fertilize the vegetable fields, Joe buys chicken manure from a brooder ranch. "We make our own compost from chicken manure, straw, and sawdust. Bacteria (commercially available compost starter) is added

Escondido grower Joe Rodriquez

to help it break down. We let it sit for three to six months, turn it a couple of times, and apply 30 yards to the acre."

The brothers switched to organic production for several reasons—strict chemical regulations, increased competition from Mexican crops, and the desire to improve the quality of the soil. "We wanted different crops and a different market—something Mexico wasn't doing."

The farm employs 25 to 35 workers year around. Crop production is highest in fall, winter, and spring. Joe says, "The market is flooded in the summer. It's hard to sell anything unless you have a product that's scarce."

Depending on the time of year and what is available, J. R. Organics sells to wholesalers, local grocery stores, distributors, and specialty produce stores. "We don't depend on just the vegetables or the flowers; we try both. If the market goes down in flowers, it won't go down in vegetables. We always have something," says Joe.

Common Production Practices For Central Coast Organic Vegetable Growers Including Harvest and Packing Methods

Crop	Planting Method	Irrigation Method	Irrigation Amount[1,2]	PEST MANAGEMENT			Nutrient Sources[5]	Harvest Method	Approximate Yield/AC (#1 Quality)[6]	Grading & Packing
				Number of Mechanical Cultivations[3]	Hand thin &/or Weed	Other				
Cabbage (Green)	Direct seed	Sprinkler	16"	3	Hand thin Hand weed 1x	Bt,[7] Insect. soap		Hand harvest	600 50 lb boxes	Field
Cauliflower	Transplant	Sprinkler	13.5"	4	Hand weed 1x	Bt,[7] Insect. soap		Hand harvest	800 12 ct boxes	Field
Cucumber	Direct seed	Sprinkler	11"	3	Hand thin Hand weed 2x			Hand harvest	900 24 lb boxes	Field
Garlic	Hand or machine plant (cloves)	Sprinkler (to bulbing stage), furrow after	11.5"	7	Hand weed 3x	Flame weed 2x	Foliar feed 3x	Windrow Contract/ Hand bag Trim, top	330 30 lb boxes	Shed
Lettuce (Leaf and Romaine)	Direct seed (coated) or transplant	Sprinkler	15"	2	Hand thin, Hand weed 1x	Bt,[7] Insect. soap	Foliar feed 1x	Hand harvest	Romaine:525 Leaf:575; both 24 ct boxes	Field
Onion (Red and Yellow)	Direct seed (coated)	Sprinkler	24"	5	Hand weed 2x	Flame weed 1x	Foliar feed 2x	Top and bag Contract/ Hand bag	Red:750 40 lb boxes; Yellow: 850 50 lb sacks	Machine sort grade & pack
Peas (snow & snap) non-staked	Direct seed	Sprinkler	10"	2	Hand weed 1x			Hand harvest	Snow: 375, Snap: 500; both 10 lb boxes	Field
Peppers (Bell)	Transplant	Furrow	24"	4	Hand weed 2x			Hand harvest	Green:800, Red:550; both 28 lb boxes	Field
Sweet Corn	Direct seed	Sprinkler (seedling stage), furrow after	24"	3		Trichogramma (for release) Phermones (for monitoring)		Hand harvest	350 48 ct boxes	Shed
Winter Squash	Direct seed	Furrow	18"	3	Hand thin, Hand weed 1x	Pyrellin E.C.		Hand harvest	Larger varieties: 800 35 lb boxes Smaller varieties: 500 20 lb boxes	Field/shed

[1] Does not include preplant irrigation.
[2] Amount will vary depending on season planted and soil type.
[3] Does not include preplant cultivation.

[4] Rodents are controlled by trapping for all crops.
[5] Compost (or manure) and gypsum are applied during land preparation for most crops.

[6] Refer to table on p. 123.
[7] *Bacillus thuringiensis* Berliner, var. kurstaki

Trade names are used to simplify information, they are not an endorsement or recommendation.

From: Klonsky, Karen, Laura Tourte, David Chaney, Pete Livingston, Richard Smith. 1993. Production Practices and Sample Costs for a Diversified Organic Vegetable Operation in the Central Coast. Davis, CA: University of California Cooperative Extension.

Organic livestock and products

The requirements for organic production of meat or poultry products are not yet as well defined as requirements for crop production. Farmers, consumers, marketers and government officials are working out agreements on these specific issues:

- The length of time the animal has to have been fed organic feed—may be different for animals to be slaughtered for meat than for poultry raised as laying stock.

- The amount of space and mobility allowed the animal, cleanliness, shelter from weather, and access to the outdoors.

- The use of synthetic vitamins and minerals and antibiotics and other medicines.

Biodynamic Farming

John Jeavons' farm, Willits

Biodynamic farming is a form of sustainable agriculture developed from the suggestions of the Austrian philosopher, Rudolf Steiner. It is an attempt to maximize soil fertility through environmentally balanced farming practices.

Biodynamic farming achieves high yields in small spaces. Generally, raised beds are used. The soil in the planting area is prepared by "double digging" and the addition of compost which increases air penetration and the capacity of the soil to absorb and hold moisture. Double digging prepares the soil two spades or about 24 inches deep. The farmer removes a spade's depth of soil and puts it aside. Organic materials are dug into the next spade's depth. Then, organic materials are mixed into the soil first removed and that mixture is put on top of the deeper mixture. With the soil prepared this way, roots can penetrate deeply into the soil.

Pleasant Grove Farms:
Converting to Organic Grain Crops

by Wynette Sills, Pleasant Grove Farm, Sutter County

In 1985 we planted our first field of organic popcorn—45 acres. Today, we have 2,000 acres devoted to organic rice, popcorn, yellow corn, oats/vetch, and almonds, wheat, and beans.

Our reasons for making this conversion were many. With conventional farming, with each passing year, we were spending more and receiving less. Our pest problems were not alleviated. We were spending a great deal of money producing crops that were already in over-supply and in return receiving below break-even market prices. The restrictions placed on conventional farming practices seemed to be escalating rapidly. We needed to make changes, and look for alternatives.

Rather than search for a "specialty" crop, we decided to grow the crops we were familiar with in a different way—a way in which our input costs would be reduced and our crop's market value would be increased. Also, we sensed an increasing consumer sentiment concerning the environment. We felt that by eliminating chemical fertilizers and pesticides our goals could be achieved.

The following strategies have helped us make this transition, and while specific to our farm, perhaps they will spark some incentive for others.

Organic Corn/Popcorn

Fertility: Rotate every other year with a legume, such as purple vetch. Because corn also needs large amounts of phosphorous and potassium, we apply composted manures every other year as well.

Weeds: Through crop rotation and precise cultivation, we have been able to suppress most weeds. However, velvetleaf remains our biggest challenge. No hand hoeing is necessary, as corn's tall stature and rapid growth allow it to tolerate weeds better than most other row crops. Johnsongrass can be a problem if it is allowed to start.

Insects and Diseases: Pests have not been a problem. Brown rot has been a problem in the almonds, greatly impacting yields. We are trying to improve the nutrient balance to create healthier trees, more tolerant of the brown rot.

Yields: Comparable to our conventional corn yields, but due to crop rotation, on any given acre, we get a corn crop only every other year.

Economics: Manure is expensive, but pre-harvest expenses are considerably lower without herbicide expenses. We can sell oats/vetch seed from the rotation crop to increase profitability. We sell our organic popcorn under our own label, trying to maximize our return. Our organic yellow corn is for human consumption or can be sold to organic livestock producers.

Organic Rice

Fertility: For our soil, a simple every other year rotation with vetch provides all the nitrogen necessary to grow a great crop of organic rice. Incorporation of crop stubble enables us to maintain soil fertility and productivity.

Weeds: We keep our water depth at six to eight inches rather than the conventional three to four inches. This suppresses most rice weeds, including barnyard grass. Crop rotation also aids in weed control. We are experimenting with no-till planting that we hope will enable us to overcome one of our worst rice weeds—roughseed bulrush.

Insects and Diseases: Neither rice water weevil nor stem rot has been a problem. Our rotation program seems to keep both of these pests in check. Continuous rice production seems to aggravate these problems.

Yields: Variable, but usually better than county average. As in corn, we only grow rice every other year, with a vetch crop in between.

Economics: Other than seed, ground preparation, and water, input costs are minimal compared to conventional rice production. We sell all of our organic rice through Lundberg Family Farms.

Plants are spaced closely and evenly in the beds. The outer leaves of the mature plants touch, creating a "living mulch" which shades the ground, keeping moisture in and decreasing weeds. The plants are watered by sprinklers or drip irrigation.

The biodynamic/French intensive method can have low start-up costs, employ soil-building techniques, use less water and fertilizer, and result in higher yields. It can be very practical for small-scale food production.

Low Capital or Low Chemical Input Farming

Low input farming means using less fertilizer, pesticide, seed, fuel, or animal feed and drugs. Purchased inputs are generally replaced with better management, increased labor, crop rotation, and soil tillage methods to fix nitrogen and control pests. Cutting inputs to lower costs can improve profits. But substituting an underutilized resource, such as manure, improves profitability only if it is equally effective and costs less than chemical fertilizer; however, it is possible that when factors in addition to fertilizer are considered, the overall benefits could outweigh the costs.

Regenerative Agriculture

The goal of farmers who practice regenerative agriculture is to improve or regenerate all parts of the agricultural system. There is no long-term degradation of the ecological agricultural system including to the people involved in it. Regenerative farming practices try to mimic nature. Diversity is increased by crop rotations, by multiple cropping, and the creation of hedgerow habitats. Organic matter is returned to the soil. Nutrient release from the breakdown of organic materials is gradual and should coincide with changing crop needs. Computer-based "expert systems" may be used to monitor soil health. In regenerative agriculture, there should be long term improvement of natural resources, including the rural community.

Holistic Resource Management

The founders of Holistic Resource Management believe the common thread that leads to lack of sustainability has been degradation of the watersheds and their ecosystems. They believe history shows that any attempt to "manage" natural resources, given the complexity of our ecosystem, must consider the entire system or assume a high risk of breakdown in the long run. There are several facets on which the practices are based:

- The sun is the natural energy resource on which we must rely.

- Human relationships must grow and be nurtured for sustainability of the land.

- The whole that you are managing must be the focus, rather than the individual parts.

■ Soils must be kept covered by plants for the stability and improvement of the land base.

■ Hoofed animals are essential for most sustainable agriculture environments.

■ A well-defined goal is needed.

Goals include fiscal stability, maintaining a high quality of life for farmers, farm workers, and consumers, monetary profit, and environmental integrity.

For more information, read Allan Savory's 1988 book, Holistic Resource Management, Island Press, Washington, DC, or contact the Center for Holistic Resource Management, 5820 Fourth Street, NW, Albuquerque, NM 87107.

Additional Resources and References

California Certified Organic Farmers, 303 Potero St., Suite 51, Santa Cruz, CA 95060. The major certifying agency in California. They sell a certification handbook and a membership directory.

California Department of Food and Agriculture, Organic Program, 1220 N Street, Room A-265, P.O. Box 942871, Sacramento, CA 94271.

Coleman, Eliot. 1989. The New Organic Grower: A Master's Manual of Tools and Techniques for the Home and Market Gardener. Chelsea, VT: Chelsea Green Publishing Co.

Edwards, Clive A., et al. 1990. Sustainable Agricultural Systems. Ankeny, IA: Soil and Water Conservation Society. 696 p.

Jeavons, John. 1974. How to Grow More Vegetables. Berkeley, CA: Ten Speed Press. 159 p. A primer on the biodynamic/French intensive method of organic horticulture.

Klonsky, Karen, et al. 1993. Cultural Practices and Sample Costs for Organic Vegetable Production in the Central Coast. University of California Giannini Foundation.

National Research Council. 1989. Alternative Agriculture. Washington, DC: National Academy Press. 426 p.

National Directory of Organic Wholesalers: A Guide to Organic Information and Resources. 1993. Davis, CA: Community Alliance with Family Farmers. 271 p.

Organic Soil Amendments and Fertilizers. 1993. Pub. 21505. Oakland, CA: ANR Publications. 36 p. A thorough introduction to composts, manures and other organic soil materials, where you can get them, how they work, and what they can do for your farm.

Pests of the Garden and Small Farm: A Grower's Guide to Using Less Pesticide. 1990. Pub. 3332. Oakland, CA: ANR Publications. 286 p.

UC Sustainable Agriculture Research and Education Program, University of California, Davis. SAREP has a newsletter and publications; they sponsor conferences and support research.

What the Farmer Needs to Know, a Summary of California and Federal Organic Production Laws. 1992. Sacramento, CA: Vaupel Associates.

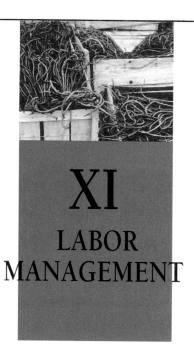

XI
LABOR MANAGEMENT

You have probably heard a great deal about the decline in worker output. Research shows that worker productivity is not a constant. As a small farmer you have much control over work outcomes. This chapter outlines some effective agricultural personnel management practices

Managing People on the Farm

Any control over production that you have on your farm is achieved through people. How these people are selected and managed affects your bottom line. People managing skills can be broken down into three essential ingredients: 1) a concern for productivity and people; 2) an understanding of labor management; and 3) purposeful action.

Concern
A concern for worker needs means attending to their well-being, as both individuals and employees. Courteous and consistent treatment, job security, fair pay, and safe working conditions are important to employees. When those needs are ignored, worker dissatisfaction may impede productivity. Developing trust between grower and employee is an important contributor to productivity and a sustainable employer-employee relationship.

Chapter Author: Gregory Encina Billikopf, Area Farm Advisor, Stanislaus County Cooperative Extension. This chapter is abridged from Labor Management in Ag: Cultivating Personnel Productivity. Available from Stanislaus County Cooperative Extension, 733 County Center III Ct., Modesto, CA 95355.

Employees expect that management will 1) value their feelings and opinions; 2) provide positive feedback for work well done; 3) meet the agreed-upon terms and conditions of employment; and 4) be consistent and courteous. Worker performance is often enhanced when employees believe they are contributing to a valuable product and are part of an effective team. How employees' needs are met has a direct bearing on their performance.

Understanding

Effective labor management demands a clear understanding of its principles and familiarity with its tools. Managers deal with a complex web of interrelated elements. For instance, the wage scale advertised may affect the quality of applicants you may recruit, and the qualifications of those ultimately hired will in turn determine the amount of on-the-job training needed.

There are a number of options available for solving people problems. If you are comfortable using only a few management tools, you may be limited in your response to a challenge. Time and effort may be required to develop or improve management competence, but they will pay off.

Howard R. Rosenberg, an agricultural labor management specialist at UC Berkeley, developed a useful model or overview of human resource management (see table). The column to the left shows the influences and constraints placed on the workplace, the center column displays the practices, decisions, and tools of labor management, and the column to the right indicates potential results.

Labor Management Influences, Practices, and Results

Influences and Constraints	Practices, Decisions, and Tools	Results
Tradition	Organizational structuring	Productivity
Competitors	Job design	• quantity
Laws	Recruitment	• quality
Labor market	Selection	Waste
Technology	Orientation	Breakdowns
Union contracts	Training and Development	Satisfaction
Individual differences	Supervision	Motivation
	Performance appraisals	Absenteeism
	Compensation	Turnover
	Benefits	Strikes
	Safety and Health	Grievances
	Organizational development	Litigation
	Research and evaluation	Injury and Illness

Adapted from H.R. Rosenberg, "Management Choices Front and Center," Labor Management Decisions, Spring, 1991.

Management tools can help temper challenges or improve results. For instance, an employer might choose to hire the first 20 applicants who show up for a cherry-picking job without testing their skills. By so doing, she foregoes the opportunity to use a selection filter to hire more productive workers.

Action

Understanding and concern without action can be like planting and cultivating without harvesting. It is not easy to confront employees with their poor performance, listen to their difficulties, act as an effective mediator or arbitrator to reduce conflict among them, or take an unpopular yet principled stand in the face of adversity. But purposeful action—carrying out a plan to obtain a specific result—may help you turn challenges into opportunities.

What hinders us from taking action or reaching goals? Two reasons, according to Vroom's Expectancy Theory, include: 1) the benefit may not seem worth the effort; and 2) we may doubt that the effort will yield the desired result. Two additional challenges may include lack of self-esteem or ability to focus. Finally, action may not be effective because of faulty planning, evaluation, or correction. This is not a suggestion that managers must always act swiftly. There is a balance to be found between premature, hasty action and having good intentions but doing nothing.

Selection

Many candidates who in good faith try to learn some skills after the job is offered to them, are never able to do so—even after taking classes in the subject. A word of caution is not to take any skill, ability, or knowledge for granted!

Sometimes a new employee is needed urgently. You may want to hire a provisional employee while you go through a more formal process. The provisional employee can apply for the regular position.

Small farmers need to first decide what they are looking for. You need not be constrained by past selection decisions. The chances of finding qualified applicants increase as the applicant pool expands. Do not be afraid to advertise the position. You can filter out unqualified applicants later.

Next, determine where in the selection process you will measure the various skills needed. For instance, you may check out operating skills through a practical test and an applicant's personality through an interview. Make a list of questions and situations for the interview and test. Design test and interview situations and questions. Throughout the process paint a clear picture of what the job will be like. Some applicants will select themselves out.

What applicants say they can do in applications and interviews and what they can do on the job are not always the same. Practical tests where candidates are asked to perform a task are invaluable. The more critical interpersonal relations are, the more critical the oral interview becomes.

Reference checking involves obtaining information about applicants from previous employers. Meeting references in person—or on the phone—is usually more productive than asking them to respond in writing. Employers are becoming more hesitant about providing references given our libel-happy society.

There are doctors and medical facilities willing to work closely with employers to develop job-related physical examinations. Legally, such tests can only be given after an employer has made a job offer. Having base measurements taken on a number of important parameters including blood cholinesterase level, hearing ability, and lifting capacity may be useful in defending possible future workers' compensation claims.

A farmer who follows such an approach can greatly increase the chances of hiring the right person for the job. Although lamentable, it is well worth starting over if you are not satisfied with any of the candidates.

It is during the orientation period when employees are most receptive to change. The way employees are treated during the orientation affects how they will perform on the job.

Appraisals

Performance appraisals can help you communicate with workers to let them know how they are performing. Appraisals can also serve as a check for the effectiveness of the selection process. Finally, appraisals can help you develop data to conduct training, make pay decisions, or take disciplinary or other management action.

Evaluations work best when workers know what is important to you. Delivering the good news about employee performance can be fun, but few find it easy to tell employees where they fall short or how they could improve. Do not wait for a formal appraisal to let employees know how they are doing. It is good to be positive without misleading employees. In Employment Discrimination Law, Barbara Schlei and Paul Grossman (1983, Bureau of National Affairs) write:

> As with olives, where a small olive may be graded "large" and the largest "super" or "colossal," the worst rating many companies give their employees on appraisals is "good." Thus, the employer might be in a position of arguing that "good" actually means "bad."

Undeserved positive performance evaluations might later help make a worker's case against the small farmer for wrongful discharge or discrimination.

A farmer or supervisor can make things easier for herself by putting more responsibility on the worker for the performance appraisal. While people normally prefer not to dwell on their weaknesses, most will prefer to point out their own shortcomings than having them pointed out. When the worker has acknowledged areas that need improvement, a change in roles can take place. Rather than an expert telling someone about his faults, the supervisor can now be an active listener, offering support and help to the worker in changing dysfunctional behaviors.

Pay

Jobs that call for creativity and personal growth may provide the best motivation of all: intrinsic rewards. Satisfaction originates from within the worker. An overemphasis on external rewards may eliminate internal motivation. Internally motivated workers may not perform as the employer would want, however. Pay can be a powerful management tool.

Workers expect that pay will: 1) cover basic living expenses, 2) keep up with inflation, 3) leave some money for savings or recreation, and 4) increase over time. Simply paying more does not motivate workers to work harder. Instead, employees must see how their performance is connected to pay.

Time-based pay

The development and management of time-based pay centers on the initial setting of wages and the giving of raises. You may set a range of steps (e.g., from $5.50 to $7.50) for a particular job. Wages may be set as a result of wage surveys (that take into account what others pay) and job evaluations (that compare job factors, such as required skills or responsibility).

Workers are usually started at the lowest step and move up the pay scale. Raises come as a result of seniority, merit, or both. If you find that in order to attract new workers you need to start them higher in the scale, perhaps the scale is too low and has not kept up with inflation.

Informal incentives

According to motivation experts, informal rewards are especially effective if they are given at uneven or unexpected intervals. The worker is kept guessing when the next acknowledgment will come. A bonus given at the same time of the year, on the other hand, becomes part of what is expected in the eyes of the worker. If the bonus is not received, workers are likely to feel dissatisfaction.

Examples of informal incentives include: a pat on the back, a sincere thank you, a $100 bill, a dinner for two at a local restaurant, or a pair of tickets to the rodeo. A specific commendation, "This is for reducing our total harvest-time machinery break downs," is more effective than "Thanks for all you do."

Incentive pay

The purpose of an incentive pay program needs to be clear and specific. Measurable goals work best. Incentives are used to either encourage superior performers to keep up the good work, or for good workers to improve their performance. Tradition is not always the best indicator of what programs will work under incentive pay. Do not be afraid to be creative. Two types of incentives that should be avoided include safety incentives and chance incentives (where luck is involved). Effective incentives link expected performance with pay. Individual incentive plans offer the clearest link between a worker's effort and pay. Group incentives can also be effective, and promote team work. They are especially useful when it is difficult to distinguish individual effort.

An effective incentive system should anticipate loopholes—or ways that workers can achieve one result at the expense of other valuable results. Incentive pay should not be solely set on how much an average worker should make in an hour. Instead, incentives are likely to be successful if they are designed so that the more the worker makes, the better off the small farmer will be. Regardless of how pay standards are set, future changes may make it easier for workers to earn more, but never make it harder. Workers will fear otherwise, that performing too well may lead employers to require more of them.

The greater the control workers have over results, the greater the total percentage of their wages that can be paid on an incentive. If you are contemplating new machinery or work methods, you may want to wait to begin the incentive program until after you have made the changes. It pays to involve workers in the development of the incentive. At the very least, workers need a chance to review and comment on the system before it is implemented. In incentives, simplicity builds trust. Some incentives may take as long as six months to a year to take hold. Regardless of the pay interval, giving workers frequent feedback on their performance is critical. Keep and analyze performance records.

Supervision

The term supervisor here indicates anyone who has responsibility for directing or facilitating the performance of one or more employees. A supervisor's effectiveness will depend on how well she can train, communicate, and delegate a job to her subordinates. Interpersonal relations and fairness also help establish a positive, sustainable relationship.

Farm advisor Pedro Illic with Fresno farmer

Training
Supervisors need to assess employees' present skills; then they can set clear learning objectives. The training itself consists of: 1) explaining and demonstrating correct task performance; 2) helping workers to perform the task under supervision; 3) allowing personnel to perform alone; 4) evaluating worker performance; and 5) coaching employees based on evaluation results. Once an employee has mastered the required performance, he can further cement his skill by coaching another.

Supervisors may want to keep

the following ideas in mind when giving explanations: 1) present only a few concepts at a time; 2) where possible divide tasks into simplified components; 3) test the workers' understanding frequently; 4) involve all workers; 5) use visual aids with oral explanations (e.g., samples of detective fruit to watch for); and 6) encourage questions. Honest praise also helps.

Communication

Messages sometimes get distorted. People may hear something different from what the person speaking intended, while the speaker sometimes takes it for granted that the message is understood. Sometimes differences in languages compound the challenge. Other times messages are distorted when they pass through a supervisor. Supervisors, for instance, should not be expected to communicate information they do not understand, or to always communicate "the bad news." Finally, providing a good listening ear is also part of a supervisor's job.

Power and delegation

One-way classification distinguishes organizational power (based on positional hierarchy) from personal power (based on personal characteristics). Responsibility and power ought to be well balanced. It is hard to hold a supervisor responsible for those he supervises without allowing him to reward superior work, or discipline lousy performance. Many supervisors feel as if they have to act with one hand tied behind their backs. At the other extreme, unchecked organizational power can lead to an even potentially more serious problem, that of abuse of power. Nothing can be more harmful to a sustainable employer-employee relationship than abuse of power. Examples of power by a supervisor may include sexual harassment, showing open favoritism to friends or family members, and dishonesty.

Empowerment, on the other hand, consists of passing on decisions to the workers who have to execute them. When delegating tasks or decisions, supervisors need to be clear on how much they are delegating. Most farm personnel have ideas on how to improve the farm enterprise. When delegating a task, build a system for checking that there has been follow-through.

Dangerous pesticide application practices

Farm Safety

A farm safety program can provide: 1) a safe working climate, 2) worker training, 3) hazard evaluation and correction measures, 4) safety committees, 5) discipline for violation of safety rules, and may also involve 6) careful employee selection, including the use of pre-employment physicals. Identify safety hazards and follow through to make corrections.

Base your safety program on the needs of your operation. Make your safety plan simple and practical—one that catches your employees' attention. Follow through with it. Some farm safety training directly focuses on teaching employees about safe practices with new work methods or equipment. A substantial part of any safety training program is to remind employees about what they already know but tend to forget when in a hurry. Telling someone to be safe does not go far in causing behavioral changes. Most of us give little thought to death and injury. And that is good; we certainly don't want to be paralyzed by fear. Nevertheless, farming can be a hazardous occupation, unless we take the necessary precautions.

Safety training in itself doesn't prevent accidents. But, if it captures the employees' attention, training can lead to changes in attitude, and in turn to changes in accident-causing behavior. Showing workers slides or videos of workers who have suffered injuries can be effective.

Short yet frequent training is generally more effective than a single, long meeting. Frequent meetings serve to remind workers of the farmer's commitment to safe practices. Seasonal meetings can be tailored to specific agricultural activities.

Discipline

At times workers do not seem to be working out. Most farmers experience some degree of discomfort when dealing with such personnel. An effective disciplinary process can help improve difficult communication with workers. Workers understand what is expected of them. When an effective disciplinary program is established, challenges are often resolved before they get out of hand. A properly constructed disciplinary policy puts the burden of improvement on the worker. A useful way of determining consequences for disciplinary violations is to ask, for every rule and consequence: What would I do if my best employee . . . didn't call in when he was sick? . . . came to work late? . . . was caught sleeping on the job? One may then be confident that the rule will not do more harm than good.

No matter how hideous the infraction, no worker ought to be terminated on the spot. Instead, the employee can be calmly asked to immediately leave the premises after being informed that he has been suspended, say, for three days. It may be necessary to drive the employee home if his driving may be unsafe. The worker is also informed that when he returns from the suspension a decision will be made as to whether he will be terminated, given further discipline, or if some other action will be taken. This allows the management team to make a more calculating decision and one less tempered by the passion of the moment. Also, in California farmers are required to give employees their final paycheck immediately upon dismissal.

When a just cause approach is followed, most workers who are terminated will agree that the employer acted in a fair manner. Excellent guidelines are offered in

Grievance Guide (1987, Bureau of National Affairs). Disciplinary policies need to be fair in substance and in application. They also need to be well communicated. Beyond that, employers have quite a bit of latitude. Some worker behaviors, although offensive to some employers, may have to be tolerated. Employees have some right to make personal choices regarding dress and grooming, especially when they are not in contact with the public. Different degrees of fault should bring different consequences. Repeat offenses can be met with more serious responses. Rules that made sense in the past but do not apply to the present operation need to be dropped.

An unenforced rule soon ceases to become a rule. If discipline enforcement has been lax in the past, employees need to be alerted to the intended change. When making a careful investigation, consider if special circumstances may have justified the actions of the employee. The more serious the accusation, the greater the proof of employee wrongdoing required. For situations involving possible criminal activity you may want to call the police. Throughout the disciplinary process the farmer must act as if truly interested in helping the worker with the problem. Farmers need to balance their need to treat workers in a non-arbitrary fashion and yet allow for special circumstances. It does not make sense, for instance, to equally punish a first time wrongdoer and a repeat offender.

Additional References and Resources

Billikopf, G. E. People in Ag: Managing Farm Personnel (free monthly newsletter). Modesto, CA: University of California Cooperative Extension.

Billikopf, G. E., and L. A. Sandoval 1991. Systematic Approach to Employee Selection (video). Davis, CA: University of California Visual Media.

XII
KEEPING THE
FAMILY FARM HEALTHY

Life on a family farm can be hectic and, at times, seem out of control. Often life is fast-paced and decisions need to be made quickly. This can keep a family farm operator from attending to some critical basic needs of the business. Setting aside time to focus on the following four areas will provide your business with a solid framework for handling day-to-day activities as well as dealing with unique events that affect your ranch or farm:

- Developing a structure to support your work.

- Acknowledging the dynamics of family business life and working with them.

- Talking about decision-making practices before you need to use them.

- Developing and maintaining an education program for all members of the business.

Developing a Structure to Support Your Work

Creating a structure that supports your efforts should include the following elements.

Chapter author: Amy Lyman, Organizational Consultant, Great Place to Work, San Francisco, CA

Job Descriptions

Develop job descriptions for all family members involved with farm work, including a definition of the role of board members. Job descriptions set the boundaries of each person's responsibility and designate individuals as the primary holders of certain information. Job descriptions are not straight jackets; they provide guidelines that define how and by whom the work will get done.

Cross-training

Creating job descriptions helps to prepare you for cross training your employees. Cross training will help you identify the information needed to run the operation, and help you figure out what areas need back-up support. Having two people trained to do the same job (although one person holds the primary responsibility) lets employees and managers know who to call if an employee is absent, instead of frantically searching for someone to fill in.

Measure Your Income and Expenses

You need a way to measure your income and expenses that is accessible and understandable to family members and employees. Too often family members consider major financial decisions without clearly understanding the impact different choices would have on the financial health of the business. Whatever system you use to track income and expenses must be accessible and understandable to all who need to use it.

Specific Times to Hold Meetings

Choose specific times to hold meetings on important issues. This is very simple, involves some pre-planning, and will save you time in the long run. Rather than calling a meeting the night before, set aside time for important meetings throughout the year. By pre-planning your meetings, you can work around peak work periods and give everyone notice so they will come prepared to make the best decisions. Setting up meeting times does not mean grounding them in cement, but it does give everyone time to prepare.

Planning for Major Events

Plan for major events, whether they be purchases of equipment, building or renovating a structure, or doing major maintenance work. There never seems to be enough time to get things done, and often it takes a crisis before something finally happens. If you use part of a scheduled meeting to brainstorm about the items that need to be taken care of, and prioritize those items, you are providing yourself with useful guidelines for the work to be carried out over the next three to six months.

Creating a Family Business Vision Statement

This document states why you are doing what you are doing and how you want to see it get done. It reflects family members' ethics and values that guide business operations. This is especially important for family-owned businesses because it clarifies why you are in a family business. Is there something about your family's history or the way you run the operation that is important to you? These family concerns can be incorporated into the vision statement and serve as a guideline for making major decisions on issues such as hiring, land use practices, and long-term growth plans. It doesn't take a tremendous amount of time to develop a vision statement; however, it takes a good deal of thought. Once you've started to talk about your vision, it is easy to write the statement.

Acknowledging the Dynamics of the Family Business

The second major area of focus concerns the unique dynamics that operate in a family business. Many management consultants tell people to separate their family lives from their business lives, and only make decisions based on one set of concerns or relationships. This is not possible in a family business. In a family business, a sound decision takes into consideration personal as well as business issues.

Paying attention to both family and business concerns is not easy. Family rules and norms about what to talk about and what to keep quiet about may limit discussion of important issues. As you begin to discuss family and business concerns, you may need to use a non-family member to help you sort through the questions that come up. But, families tend to be closed groups. This can affect the family's willingness to call on outside resources for help. If you think, "We can do it ourselves," or "We don't need any outside help," remember that one of the advantages outsiders have is they're not caught up in your rules and norms. They can ask the "dumb" questions that will trigger solutions to the problem that had been stuck or hidden.

A valuable aspect of family involvement in the business is that you can call on reserves of loyalty and commitment not available to public companies. A commitment to see the operation succeed is very powerful and can see you through hard times with much less pain than a public company might experience. That's a very positive aspect of the family business.

Another area of family dynamics concerns traditional family roles. More women, especially daughters, are going into family businesses and moving into management positions or positions of non-traditional work responsibility. We don't need to stick with traditional ideas about what is appropriate for a woman or a man to do. If you are looking for a family member who is going to be the next person in charge of personnel or machinery or crop production, look at individual ability. Are you training both your sons and daughters to take over the farm? Who is interested in the operation?

Sometimes a daughter rather than, or in addition to, a son is interested in joining the farm. Don't overlook your daughter because of her gender. We have all seen many examples of women and men taking on non-traditional responsibilities. The time you usually discover someone's true capabilities is during a crisis when you need help and it doesn't matter if it's a non-traditional job or not. You find that women can drive tractors or milk the cows and men can keep the books. It is important to look at all family members and their abilities. If the ability is there, the fact that they're sons or daughters becomes irrelevant.

To make the most of your family business, create and use a diverse network of people resources for both your business and family concerns. Bring all family members and their resources into the business so you don't just use the resources of the founder or first generation. This can be hard when the resources that the younger generation is bringing in are new or unfamiliar. An example is the use of computers and computer programs for bookkeeping or keeping records on the health of the herd. There's always a little bit of anxiety when a junior family member knows how to run the computer better than a senior family member. But, the advances that people can take advantage of now are the same kinds of advances that senior family members were able to take advantage of in earlier years. Each family member brings with him or her a unique network of resources. Use all of them.

Talking about Decision-Making Practices before You Need Them

How do you make decisions? Is everyone called together on a certain day to make an immediate decision? Do people feel uncomfortable because they don't have all the information they need and haven't had time to think through various courses of action?

Farm advisor Pedro Illic with Fresno farmer

There are many things you can do to prepare yourself to make decisions. Talk about the importance of being honest instead of saying what people want to hear. When it's your operation and you're the ones making the decisions, you need to have people be honest about their anxieties and questions. All who are involved in the decision should be encouraged to speak up about what they perceive as the pluses and minuses of decisions and their willingness to take collective responsibility.

Collective responsibility means that everyone agrees to be responsible for the decisions that are made. This brings with it a tremendous amount of individual responsibility and comes in handy as a philosophy when there is dissension within the family. At times, family members who don't get along pay less attention to decisions made that affect someone else's work area. They accept decisions that they are not comfortable with because they don't see them as their decisions. Collective responsibility means everyone who is involved in the operation shares responsibility for the decision. It's a very powerful tool in terms of encouraging people to speak up.

Also, consider the relationship between trust and risk. Trust involves risk. Part of risk is supporting someone's efforts to take on a challenge. When you trust someone you accept the risk that some things will be out of your hands. There may be times when you feel uncomfortable because a decision has been made but someone else is responsible for doing the job. The risk is letting go of the anxiety and saying, "I trust you. We made this decision. It's your area to implement. I trust you to do it and to do the best that you can." This process does not imply taking unnecessary risks by making a decision with which you are not comfortable. The focus should be on trusting the person to do his or her best once the decision has been made.

When decisions are made, pay attention to both family and business concerns. It's very important to consider the family reasons and the family dynamics that are influencing the business decisions. Often in family businesses, decisions are made ignoring the family background that influences the decision. Sometimes it's hard to separate the two. Calling in an outsider may help. Business decisions made in a family business have a family component. Sometimes it's 90 percent family concerns and 10 percent business concerns and sometimes it's the other way around. For example, consider a family that owned a large parcel of land that was split up, with parts of it sold to different farms over time. Now the family wants to buy the land back to recreate the family homestead. If you talk about it in terms of business decisions, you can get into all kinds of convoluted justifications for why you want to buy the land. Acknowledging that it is the family homestead and that's why you want to buy it back frees you from trying to justify something that no one really wants to talk about. A decision can be made based on the impact of the decision on the business and the family. Can the business support this purchase? Is it going to hurt the operation? Will it help the operation? Are you going to feel good about it?

Decision making involves responsibility as well as opportunity and can be affected by an individual's family and business roles. To successfully make decisions, the responsibility that comes along with all of the opportunities needs to be emphasized. Accepting responsibility for business decisions can be a difficult area for board members who are not actively involved in business operations but participate in decision making. They can have a hard time dealing with the conflicts between perceived family and business responsibilities. Without a job description, especially one that spells out board responsibilities, all a family member has to rely on are family responsibilities, and that's only half the picture. All family members who are involved in decision making should have job descriptions that cover their business responsibilities. This makes clear what their obligations are to the business, and what the boundaries are around the opportunities that can be pursued.

Developing Opportunities for All Employees to Participate In Educational Programs

Developing specific educational programs for family members coming into the business is especially important to insure that knowledge is transferred visibly rather than assumed to be learned. The next generation family member may have worked on the farm during the summer or helped out in one area, which is a good start but does not provide an understanding of the total operation. The founder or entrepreneur has done it all and knows everything. Work got done because there was no one else to do it. The next generation is in a very different situation. Developing an education program helps them understand their responsibilities and develop a greater degree of confidence in their abilities. It also insures that other employees see that knowledge is consciously transferred.

Educational programs that all employees can participate in are a good enticement for attracting the best people to your operation. Including opportunities for education as a part of the job says that you care not only about what the employee can do now but about what he or she will be able to do in the future. This might not work for every employee, but for your key employees it is very important to state that you are going to help them further their education.

Education for career advancement may involve taking time off from the farm. One issue that often comes up for next generation family members is the dilemma of starting work on the farm or ranch right after high school. Questions such as "Can I do anything else?" " Am I just here because I'm a family member?" "Do I really have skills that are transferable?" pop-up after a few years.

It is often recommended that next generation family members work somewhere else for two years prior to entering the family business. There are a number of reasons for this. One is to gain a sense of independence and a sense of confidence in their own skills. Another is to change the relationship between parent and child to an adult-adult relationship. Education for career advancement and support for pursuing educational opportunities away from the farm are very important to the development of the next generation of managers.

One of the most difficult family business issues concerns retirement. Who will be the next leader? It may be difficult for a retiring farmer and his or her family to talk about what's next. He or she may have built the operation or worked on it all of his or her life. Placing attention on what a person will do after he or she is finished farming implies a beginning of another lifestage rather than an end to an active life.

Perhaps the next stage will be teaching, working in the community or working with other small farmers who are struggling to get started. There is a wealth of knowledge, skill, and wisdom in people about to retire that could be passed on. We don't tap into this resource very well. One program called SCORE—the Service Corps of Retired Executives—makes connections between retired executives and other organizations where there is a need for part-time help in areas such as accounting, production processes, or marketing. You can do the same thing in terms of a farming operation. Think about what's next!

Additional Resources and References

Canadian Association of Family Enterprise, 121 Bloor Street East, Suite 803, Toronto, Ontario, M4W 3M5, Canada.

Community Alliance with Family Farmers, P.O. Box 464, Davis, CA 95617; (916) 756-8518.

Dyer, W. Gibb. 1986. Cultural Change in Family Firms: Anticipating and Managing Business and Family Transitions. San Francisco, CA: Jossey-Bass, Inc.

Errington, Andrew (editor). 1986. The Farm as a Family Business: An Annotated Bibliography. Reading, England: University of Reading, Department of Agriculture.

Family Firm Institute, 12 Harris Street, Brookline, MA 02146; (617) 738-1591.

Gamble, Dennis, Lesley White, and Stephen Blunden. A Systematic Approach to Exploring the Complexity of Family Farm Transfers. Richmond, Australia: University of Western Sydney—Hawkesbury.

Gasson, Ruth, et al. 1988. The Farm as a Family Business: A Review. Journal of Agricultural Economics: 39(1):1-42.

Keating, Norah C., and Brenda Munro. April, 1989. Transferring the Family Farm: Process and Implications. Family Relations: 215-218.

Personnel Decisions in the Family Farm Business. 1993. Pub. 3357. Oakland, CA: ANR Publications. 64 p.

Thomas, Kenneth, and Bernard Ervin. 1989. Farm Personnel Management. Pub. 329. St. Paul, MN: University of Minnesota, North Central Regional Extension.

Ward, John L. 1987. Keeping the Family Business Healthy: How to Plan for Continuing Growth, Profitability, and Family Leadership. San Francisco, CA: Jossey-Bass, Inc.

Weigel, Daniel J., and Randy R. Weigel. 1987. Keeping Peace on the Farm—Two-Generation Farm Families. Pm. 1292. Ames, IA: Iowa State University, Cooperative Extension Service.

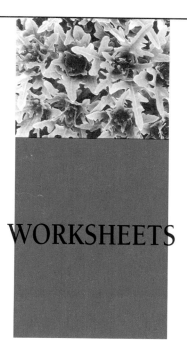

WORKSHEETS

Worksheet compilers: Daniel W. Block, Professor, California State Polytechnic University, San Luis Obispo; Fred Conte, Aquaculture Specialist, Animal Science Extension, UC Davis; Shirley Humphrey, Staff Research Associate, Small Farm Center, UC Davis; Eric Mussen, Extension Apiculturist, Entomology Extension, UC Davis; Steve Sutter, Farm Advisor, Fresno County Cooperative Extension

Business Qualities Checklist

Each family member should complete a copy of this worksheet. Analysis of the results can provide insight into the areas of strength of various family members. Check as many statements as describe your personal qualities.

1. **Drive**
   _____I am curious and enjoy taking a chance.
   _____I am enthusiastic and a self-starter.
   _____I have plenty of energy and vitality.
   _____I persevere and finish what I start.
   _____I have self-confidence and the courage to take risks.
   _____I am able to and like to work on my own.
   _____I enjoy working hard on my own projects.

2. **Clarity of thought**
   _____I am organized and have a thorough plan of action.
   _____I seek advice for areas in which I lack knowledge and expertise.
   _____I am capable of making good decisions in a short period of time.
   _____I have the ability to acquire knowledge about business and finance.
   _____I can perform different tasks simultaneously.
   _____I have self-discipline and can keep cool under pressure.

3. **Imagination**
   _____I often think of innovative ways to solve problems.
   _____I have creative ideas relating to business.
   _____I enjoy doing things differently.

4. **Leadership**
   _____I work well with people.
   _____I genuinely like people.
   _____I enjoy supervising other people.
   _____I am responsible and enjoy providing leadership.
   _____I keep pledges and other people's trust.

5. **Experience**
   _____I have a high degree of proficiency in a technical area.
   _____I have managed a business before.
   _____I keep good records of production and finances.
   _____I know how to interpret financial statements

How did you do? The more check marks you have the better. Only an individual with exceptional business management potential will check all the qualities. If you placed a few checks in each section, you possess a good set of skills with which to approach farming. If you currently operate a farm, you should have checked several items. If you did not, think about whether or not an independent farm-based business is really right for you. If you are thinking about leaving a full- or part-time job to start farming, you may or may not have checked numerous items. If you possess very few personal business qualities, ask yourself if an independent farm-based business is right for you. If some other family member possesses more of the personal qualities that lead to success in business, can you delegate management of the farm to that person?

From: Farming Alternatives—A Guide to Evaluating the Feasibility of New Farm-Based Enterprises. 1988. Ithaca, NY: Northeast Regional Agricultural Engineering Service, Cornell University.

Family Needs and Preferences

Read each statement and check the appropriate column to rate the extent to which you agree with the statement. Each member of the family should complete this exercise independently on a separate copy of the worksheet.

	Agree strongly	Agree	Disagree	Disagree strongly

Section 1.
Satisfaction with Current Employment

1. Our family spends too little time together.
2. We usually have enough spending money.
3. I would rather spend less time working on an off-farm job.
4. The farm business is too much to handle now.
5. When our family gets together, we spend too much time talking business.
6. When we're busy, friendliness in the family disappears.
7. I like my present role in the farm just the way it is.
8. Family relations are strained because there isn't enough money.
9. I too often sacrifice things I like to do for the sake of my employment.
10. The current farm business doesn't hold my interest or use my skills very well.

Section 2.
Willingness to Take Risks with New Enterprises

1. I wouldn't mind if our new enterprise took a few risks to make more money.
2. If the new enterprise fails, I'd like to try another.
3. When there's a chance we'll lose money, I'm tense.
4. I would risk losing our savings on a new enterprise.
5. I would be willing to borrow money to finance the new enterprise.
6. I prefer a low risk/low profit venture to a high risk/high profit venture.
7. If our new enterprise fails, I'd consider getting out of farming.

Family Needs and Preferences

	Agree strongly	Agree	Disagree	Disagree strongly

Section 3.
Enterprise Preferences

1. I like to work with livestock.
2. I like to work with crops.
3. I like to work with machinery.
4. I like to do a variety of farm tasks.
5. I like to have a steady workload over the year.
6. I like to work very hard for a few months, then take a few months of rest.
7. Supervising workers is unpleasant.
8. I could easily supervise more workers.
9. I enjoy meeting different kinds of people and making them feel "at home."
10. Having people drop by unexpectedly and at odd hours annoys me.
11. I would be good at sales because I enjoy seeing a satisfied customer.
12. The new business must use my talents and skills more than the present one does.
13. The most important quality that a new enterprise should have is the ability to make a good profit.
14. It's important for our business to be unique.
15. I would enjoy an off-farm job.

From: *Farming Alternatives—A Guide to Evaluating the Feasibility of New Farm-Based Enterprises.* 1988. Ithaca, NY: Northeast Regional Agricultural Engineering Service, Cornell University.

Interpreting your scores

All members of your family will be affected by occupation or enterprise changes. It takes an immense amount of time to get an operation going. That time takes away from family needs. Most small farms are managed and operated by families and the farm business usually shares space with household space. To avoid family resentment, a new business must satisfy more of the family's needs and lifestyle preferences than does the present business or job.

After each family member has completed the questionnaire, identify the strengths you would bring with you as you enter farming. It's important for you as a family to decide which issues are of low interest or present potential problems. Do your responses suggest unity and prospects of teamwork? Are answers similar in every category? If so, the family is exceptional and will experience fewer difficulties than most families. The average family will disagree on many subjects but will also have areas of agreement. Are you all willing to risk losing the entire farm business? Should you limit risk to a set dollar investment on a trial basis? What role does each family member wish to play in the farm? If your family doesn't work well together, you will not work well together in the business.

Section 1.
Satisfaction with Current Employment
Responses in the first section indicate your level of satisfaction with your present occupations. In family-operated businesses, the principal manager often has a more positive attitude toward the business than other family members. Look for areas in which family members are very satisfied or dissatisfied.

Section 2.Willingness to Take Risks with New Enterprises
Changing employment involves taking new risks. Is everyone equally comfortable taking risks on the new venture? If two or more people with financial stakes in the business feel radically different about risk, conflict may occur. The family should agree on a mutually acceptable level of risk.

Section 3.Enterprise Preferences
The first four questions identify the types of commodities with which family members like to work. Questions 5 and 6 identify preferred work pace. Year-around enterprises are less hectic than those that specialize in seasonal commodities like Christmas trees. Questions 7 and 8 concern labor management. If employees will be hired, someone will need to recruit, train, and supervise them. Is someone in the family willing to handle supervision? Questions 9, 10, and 11 help you assess your willingness to be in the people business. If customer service is involved, one of the workers must be outgoing and enjoy dealing with people. The last four questions identify features that family members would like in the new enterprise.

Climate Requirements			
Required for operation		**Provided by your site**	
Days to harvest		Average date of last spring frost	
Lowest temperatures tolerated (may need greenhouse)		Average date of first fall frost	
Highest temperatures tolerated		Days in growing season	
Average annual rainfall		Expected winter low temperature	
		Expected summer high temperature	
List anticipated climate problems.			
List possible solutions and estimated costs.			

From: Farming Alternatives—A Guide to Evaluating the Feasibility of New Farm-Based Enterprises. 1988. Ithaca, NY: Northeast Regional Agricultural Engineering Service, Cornell University.

Once you complete the data in the table, you will be able to determine which crops can and cannot be grown with your climatic conditions.

Soil Requirements		
	Required for enterprise	**Characteristics of your site**
pH		
Moisture/drainage		
Fertility		
Topography		
Acreage		
Other		
List anticipated soil problems.		
List possible solutions and estimated costs, such as importing soil to raised beds.		

From: Farming Alternatives—A Guide to Evaluating the Feasibility of New Farm-Based Enterprises. 1988. Ithaca, NY: Northeast Regional Agricultural Engineering Service, Cornell University.

After completing the worksheet, compare the needs of the crop with the soil characteristics and note potential limitations. Some "poor" soil conditions may be just right for alternative crops: blueberries in low pH soils, watercress on flooded soils or creek beds, and some herbs in low fertility soils.

Consider the condition of the soil if it is to support structures, ponds, roads, septic systems, and parking lots.

Water Requirements

Is this going to be a dryland farm?

    Yes _____   No _____

If not, what type of irrigation do you intend to use?

    Furrow _____
    Sprinkler _____
    Drip _____
    Other _____

Where will your water originate?

    Well _____   State water project _____
    Federal water project_____   Other _____

Is water quality appropriate?

    Yes _____   No _____

Is water quantity adequate all year?

    Yes _____   No _____

From: Farming Alternatives—A Guide to Evaluating the Feasibility of New Farm-Based Enterprises. 1988. Ithaca, NY: Northeast Regional Agricultural Engineering Service, Cornell University.

Machinery and Equipment Requirements

Machinery and equipment	Size or capacity	Own	Purchase	Lease or rent	Estimated cost

From: Farming Alternatives—A Guide to Evaluating the Feasibility of New Farm-Based Enterprises. 1988. Ithaca, NY: Northeast Regional Agricultural Engineering Service, Cornell University.

Farm Structures

What buildings are on the property and what is their condition?

Do you have structurally sound fences?

    Yes _____   No _____

If you need additional buildings or fences, what is their cost?

From: Farming Alternatives—A Guide to Evaluating the Feasibility of New Farm-Based Enterprises. 1988. Ithaca, NY: Northeast Regional Agricultural Engineering Service, Cornell University.

Enterprise Selection

General questions

Is the enterprise adaptable to your area?

Yes _____ No _____

What is your experience with the enterprise? _____

What research has been done? _____

Where else is the enterprise established? _____

Is acreage increasing or decreasing? _____

Information access

Are you familiar with your County Cooperative Extension farm advisor, Farm Bureau personnel, USDA officials, and your local library?

Yes _____ No _____

Is sufficient information available?

Yes _____ No _____

Are you willing to learn new skills?

Yes _____ No _____

Profitability

What are the total production costs? _____

What kinds of yields can you expect? _____

What is the expected gross and net income? _____

What variation in net income can you expect? _____

How does this crop compare to other crops? How does this crop compare to livestock? _____

_____

Labor

Have you checked the regulations of the California Labor Law?

Yes _____ No _____

How many acres of the crop can you manage with the labor you have? _____

Would it be economical to buy or rent labor-saving systems—mechanical transplanter, selective herbicides, larger vehicles?

Yes _____ No _____

Enterprise Selection

Is seasonal labor available?

Yes _____   No _____

What will be your monthly labor needs? _____

Are you planning to use family or hired labor or both?

Family _____Hired _____Both_____.

Have you considered the cost of your own labor?

Yes _____   No _____

Financial factors

How much capital will you invest? _____

Will you borrow capital?

Yes _____   No _____

What will be the terms of loans (length, interest, equity requirement, collateral)?

_____

Is a high rate of return on your investment important?

Yes _____   No _____

Will you consider risky enterprises?

Yes _____   No _____

Rotation considerations

How does the crop fit into rotation with other crops planned? _____

_____

How long does it take to mature—from planting through harvest? _____

_____

What will be the effect of weed management practices used on the farm? _____

_____

Is the crop susceptible to soil-borne diseases?

Yes _____   No _____

Pollination

Which trees, rows or field crops require pollination? _____

_____

How many hives will be needed? _____

Enterprise Selection

How will you determine hive quality? _____

_____

What will the rental cost be? _____

_____

Pest management

What are the pest problems? (Historically important pests are known by the local farm advisors.)

_____

Are there pesticides for the crop registered in California? _____

_____

Are there varieties available that are resistant to disease? Do they have good yields? Are they high quality?

_____

Harvesting

How many crops will you grow? _____

_____

How many harvests of each crop are required to make money? _____

_____

How is the harvest interval affected by temperature? _____

_____

How long will it take to harvest? How will the crop be packaged for market? What will containers cost?

_____

Do you have cooling facilities for perishable products?

Yes _____   No _____

Marketing

Are you familiar with market quality standards for the crop?

Yes _____   No _____

Have you studied the market history and trends of the crop? (Don't base crop selection on recent high market prices.)

Yes _____   No _____

Enterprise Selection

What marketing methods will you use: conventional markets including wholesalers, brokers, retail outlets, cooperatives, or contracts with processors; or direct markets including farmers' markets, roadside stand, and U-picks? _____

_____

How close are you to potential markets? _____

_____

How much time are you willing to spend marketing your products? _____

_____

Are you familiar with marketing regulations?

Yes _____   No _____

Adapted from Klonsky, Karen, and Patricia Allen. Enterprise Selection. Davis, CA: Small Farm Center, UC Davis.

Management Structure

List the person responsible for each of the following jobs and estimate the percentage of time that will be spent on it. Also write a short job description for each job.

Production/farming operations

Person responsible _____ % time _____

Short job description _____

Marketing/sales

Person responsible _____ % time _____

Short job description _____

Finance/administration

Person responsible _____ % time _____

Short job description _____

Compiled by Shirley Humphrey, Staff Research Associate, Small Farm Center, UC Davis and Eric Mussen, Extension Apiculturist, UC Davis

Marketing

Target market description (identification of potential customers)

Demographics:
Age range, income level, geographic location of residence or work place, number of children, marital status, ethnic group, education level, etc.

_____
_____
_____
_____
_____

Expectations:
What do buyers want and expect from your product or service?

_____
_____
_____
_____
_____

Marketing Options
Based on your research, what are the most promising markets for your enterprise?

Market 1 _____

Market 2 _____

Market 3 _____

Existing Market Demand
Number of potential buyers included in your target market at this time _____

Average purchase volume (or frequency of service) per buyer per year in target market _____

Total purchase volume (or number of services) per year in your target market _____

(Multiply the number of buyers by the average volume or frequency.) _____

Competition
What type of enterprise do you plan to develop? _____

Who are your two to three major competitors? _____

What are the major strengths and weaknesses of competitors' enterprises? _____

Marketing

Market trends

Describe the following marketing trends over the last five to 10 years of your product or service and then project future trends for the next five to 10 years:

Per capita consumption (or production) _____

Prices _____

Competition _____

What are your projections for market trends in the next five to 10 years? _____

What are the underlying reasons that account for these past and future market trends? _____
_____

What sources of information did you use to answer the above questions? _____
_____

Expected price:

Based on your knowledge of past, present, and projected prices, answer the following questions:

What is the lowest price you are likely to receive in the near future? _____

What conditions would create this low price situation? _____
_____

What is the highest price you are likely to receive in the near future? _____

What conditions would create this high price situation? _____

Expected price (price you are most likely to receive in an average future year): _____

Expected Sales Volume:

Based on your knowledge of past, present, and projected market trends, answer the following:

What is the minimum volume of product (number of units) you could sell in a poor year? _____

What production and market conditions would create this situation? _____
_____

What maximum volume of product (number of units) do you believe you could sell in a good year? _____

What production and market conditions would create this situation? _____
_____

Compiled by Daniel W. Block, Professor, California Polytechnic State University, San Luis Obispo, Shirley Humphrey, Staff Research Associate, Small Farm Center, UC Davis, and Eric Mussen, Extension Apiculturist, UC Davis

Evaluating Land and Water Resources for Freshwater Aquaculture

When evaluating a site for freshwater aquaculture, look at the total land and water resource picture. Consider how a new operation would fit in with other farming or industrial activities that share the resources. Remember: a site that is to be your sole source of income must be more carefully evaluated than a site that will be part of an integrated operation where cash flow is generated from several commodities.

If you are interested in investing in an aquaculture project, ask yourself these questions:

1. What is the objective of the venture: recreational fishing, aesthetics, commercial fee fishing, or commercial production?
2. Is the site to be exclusively an aquaculture venture? What other activities might occur on the property?
3. Will aquaculture be integrated with other agriculture activities; if so, what are they?
4. What is the source of the water—reservoir, pumped well, artesian well, or running stream?
5. How many water sources are on the property? What is the capacity, volume, or flow rate of each? Does it vary seasonally?
6. What is the temperature of the water source and the annual variation and range?
7. Is it possible to secure the necessary water permits for the needed volume of water, or is a permit necessary? Is a water discharge permit needed? Will the discharged water impact a natural body of water such as a stream, river, or wetland? Will the discharge impact other private, state, or federal properties?
8. Has a chemical water test been conducted on the water source to assess its compatibility with fish culture?
9. What pumped wells are on the property? What are their volumes, depth, and location? What are their draw-down rates?
10. What is the distance between the water source and the proposed facility site? What is the elevation and fall between the water source and the prospective production site?
11. What other use demands have to be considered for the water source? Is the aquaculture operation to receive first-use water? Can the discharge water be used for other agriculture activities and, if so, what?
12. Where is the land located? How much land is available for the operation?
13. What are the elevation of the land and the associated climatic conditions, such as annual temperature fluctuation and snowfall? Is the prospective site subject to flooding?
14. How close is the property to public roads, and what is the carrying capacity of the public roads and the roads on the property seasonally?
15. Does the property allow for on-site, live-in management and allow observation of the facility from the living area?
16. Is electricity available at the production site? What are the electrical rates in that area? Can agricultural rates be obtained?
17. What buildings are on the site (houses, barns, storage facilities, etc.)? Can they be shared with other agricultural pursuits?
18. Are ponds, raceways, or tanks being considered? If ponds are planned, what is the soil profile of the site? Get a soil analysis of the site based on a representative core sample.
19. Are other activities on or adjacent to the property using agricultural chemicals? If so, what chemicals and how are they applied?
20. Have agricultural crops been grown or other activities been conducted that may have resulted in toxic chemical accumulation in the soil? If so, what chemicals are potentially present?

Adapted From Conte, Fred, 1990, Evaluating Resources, ASAQ-C8/10/90-4/92, Dept. of Animal Science, UC Davis.

Injury and Illness Prevention Plan

It is the policy of _____ to provide safe working conditions for
(name of corporation)
all employees. Our safety program's success depends on everyone's help. _____
(name)
will have authority and responsibility for maintaining the injury and illness prevention program and will be
accountable for safety practices, safety education and training, communicating safety information, and fire
protection.

Supervisors' Responsibilities

Communication and training on new processes, new procedures, new equipment, safety activities, hazards,
and safe work practices will be done by one or a combination of the following:

1. One on one conferences between the supervisor and employees.

2. Training sessions.

3. Postings on the company bulletin board, or enclosures with the employees' pay checks.

Supervisors are responsible for getting first aid, medical care, and for filling out all necessary medical forms,
Occupational Safety and Health forms, and Workers' Compensation Employer's Report of Industrial Injury
forms.

We have procedures to investigate occupational injury, illness or exposure to hazardous substances including
gathering pertinent data and making an objective evaluation of facts, statements and related information, all
of which lead to a plan to prevent recurrence. Investigation procedures include completion of a "Supervisor's
Report of Accident" and an "Employee's Report of Accident."

Procedures for correcting unsafe or unhealthy conditions and work practices will consist of one or a combination of the following:

1. Abatement.

2. Safe guarding.

3. Personal protective equipment.

4. Training.

An occupational health and safety training program to instruct employees in general safe and healthy work
practices and to provide specific instruction with respect to hazards specific to each employee's job will be
provided to all new employees and to all employees given new job assignments. Employees will be trained
when new substances are introduced to the workplace and represent a new hazard, and when the employer
receives notification of a new or previously unrecognized hazard.

Supervisors will be knowledgeable of the safety and health hazards to which employees under their immediate direction and control may be exposed.

Supervisors, under management's direction, will meet at least quarterly to talk over results of work site
inspections, employee safety suggestions, safety problems and accidents that have happened. Supervisory
employees will conduct safety meetings with their crew at least every 10 working days.

Employees' Responsibilities

Each employee has to learn and obey safety practices and rules and use all proper safety devices and protective gear. Disciplinary actions will be taken to assure that employees comply with safe and healthy work
practices.

Employees must inform their supervisor of all safety, health, and fire hazards upon discovery. Employees will

Injury and Illness Prevention Plan

not be dismissed or discriminated against for informing supervisors or owners about work site hazards. If in doubt about a health or safety matter, employees have a duty to speak promptly with their direct supervisor.

Employees must report immediately any accidents to their supervisor.

Nearest Medical Assistance

First aid supplies are easily accessible at ———————————————————— . Locations of (location)
the nearest doctor and medical facility are posted on the bulletin board(s). In the event of death or critical injury, ———————————————————— must be notified immediately. If he (she) cannot be (name)
reached, ———————————————————— should be notified. ———————————————————— (name) (name)
must report the injury or death by telegram within 24 hours to the proper government officials and to

———————————————————— .
(company's workers' compensation carrier)

Hazard Assessment and Control

Supervisors will conduct inspections to identify and evaluate unsafe conditions and work practices. These inspections will take place whenever new substances, processes, and procedures, or equipment are introduced to the workplace and represent a new occupational safety and health hazard; and whenever the employer receives notification of a new or previously unrecognized hazard. These inspections are in addition to the everyday safety and health checks that are a part of the routine duties of supervisors.

A written safety policy will be provided to all supervisory employees who will have it easily available.

Request an order form for a written (English/Spanish) model safety program, additional forms for documentation (including accident investigation), Cal/OSHA posters, catalog of safety materials, and selected CAL/OSHA safety orders from Steve Sutter, Farm Advisor, Fresno County UCCE, 1720 South Maple Avenue, Fresno, CA 93702; (209) 488-3285. There is a nominal charge for some material.

Form prepared by Steve Sutter, Fresno Co. Cooperative Extension.

Index